Student Study Guide

to accompany

Anatomy & Physiology

The Unity of Form and Function

Kenneth S. Saladin
Georgia College and State University

Prepared by
Jacqueline A. Homan

STUDENT STUDY GUIDE TO ACCOMPANY
ANATOMY & PHYSIOLOGY: THE UNITY OF FORM AND FUNCTION

Published by McGraw-Hill Education, 2 Penn Plaza, New York, NY 10121. Copyright © 2012 by McGraw-Hill Education. All rights reserved.
Printed in the United States of America. Previous Editions © 2010, 2007, and 2004. No part of this publication may be reproduced or distributed
in any form or by any means, or stored in a database or retrieval system, without the prior written consent of McGraw-Hill Education, including,
but not limited to, in any network or other electronic storage or transmission, or broadcast for distance learning.

Some ancillaries, including electronic and print components, may not be available to customers outside the United States.

This book is printed on acid-free paper.

5 6 7 8 9 0 QVS/QVS 16 15 14

ISBN: 978-0-07-735117-5
MHID: 0-07-735117-7

www.mhhe.com

Table of Contents

Part I
Survival Skills for the A&P Student

Introduction

A course in anatomy and physiology (A&P) is typically a prerequisite for admission to schools of nursing, physical therapy, and many other allied health fields. Its purpose is not only to teach you the facts of human form and function but also to train your mind to work efficiently, think logically, and assimilate technical information. Ideally, it stimulates a maturing process that will help you develop the habits, attitudes, and skills needed to succeed in your clinical courses. Acquiring these skills can also make you happier with yourself—with your productivity, your understanding, your efficiency, and your grades.

But like all worthwhile rewards in life, mastering human A&P takes effort, concentration, and dedication. The human body is an incredibly intricate device, and it takes more than a little work to understand its structure and physiological mechanisms. However hard a textbook author or classroom instructor may work to present the material clearly, the ultimate responsibility for learning rests with you. **Learning is not imparted by books, study guides, the Internet, or teachers; it is an endogenous process undertaken by you, the student.** It requires sacrifice—perhaps forgoing television or socializing—if you want to succeed. "Survival Skills for the A&P Student" is meant to give you some pointers on how to do well in this subject. However, it is up to YOU to arrive at a plan of action in order to master A&P.

Time Management

Many A&P students are in the first year or two of college, still making the transition from high school. It's not an easy transition to make. In high school your days were probably quite regimented. Except for a lunch or study period, you probably had a class every period and migrated from classroom to classroom, like clockwork, from early morning to mid-afternoon. In college you have a more self-made schedule; you may have only three classes or so each day. It *seems* as if you have an enormous amount of free time, and there are endless temptations on which to spend it. Probably the most common academic problem among beginning college students, and the greatest cause of failure, is using time ineffectively.

In reality, you *don't* have more free time in college than in high school—you have less. **You are expected to work much harder and to learn much more than you ever have before.** If you're carrying a full course load, you're expected to treat college as a full-time job, to take responsibility for your own time, to spend it as carefully as you spend your money. But all too often, this doesn't happen.

How Time Slips Away: We tend to let enormous amounts of time slip away from us, mainly in three ways: (1) procrastination, putting off chores simply because we don't feel in the mood to do them right away; (2) distraction, getting sidetracked by the endless variety of other things that seem easier or more fun to do, often not realizing how much time they eat up; and (3) underestimating the value of small bits of time, thinking it's not worth doing any work because we have something else to do or somewhere else to be in 20 minutes or so. Consider a typical day for a hypothetical student, Janet:

8:40 AM　Janet wakes up late, having watched a movie on TV until 2:00AM. She rushes around her dorm room, getting dressed and gathering her books, and has no time to eat. She gets to her 9:00 class five minutes late, and her professor watches with a look of irritation as she crosses the room to take her seat.

10:50 AM   After her 9:00 and 10:00 classes, Janet goes back to her room to study for an hour before lunch. As she walks in, she tosses her books on the bed, and switches on the TV through force of habit; soon she's absorbed in a game show. Sitting in front of the TV, she flips through her A&P book and highlights a few terms.

12:00 PM   By 12:00, she's covered three pages of the book but remembers little of it. She turns off the TV, goes to the cafeteria, and chats with her usual lunchtime friends until 1:30.

1:30 PM   Janet rushes back to her room, brushes her hair, puts on some fresh makeup, and runs off to her 2:00 chemistry lab. When she gets there, she finds she was in such a rush to get to class that she forgot to bring her lab manual, notebook, or anything to write with. She borrows a pen and reads from someone else's lab manual.

3:50 PM   Janet goes back to the dorm, copies the data into her lab notebook, and starts to read an English assignment. The phone rings, she talks to her mom until 4:45, and tells her goodbye when a friend from down the hall knocks at her door. She and her friend chat until 5:30, complaining about all the work they have to do and how each professor must think his course is the only one they're taking. At 5:30, they go to dinner together.

7:30 PM   Janet returns from dinner to study but decides she'll just quickly log on to her favorite Internet chat room and see if any of her friends are there.

10:00 PM   She tells her online friends she has to study and logs off. She works 6 of the 20 chemistry problems that are due in two days but has trouble thinking through the problems and decides it would be easier to read some of the novel assigned in English. She lies down on the bed to read.

10:40 PM   Janet falls asleep on the bed, fully dressed, until her roommate comes in noisily at 11:30. Janet decides she can finish the chemistry problems tomorrow, gets undressed, and goes back to bed for the night.

This pattern repeats itself day after day, and suddenly it's time for midterm exams. Janet is shocked to find there are only four days left until her Monday midterms in English and chemistry, and her A&P midterm is on Wednesday. She's three chapters behind in chemistry, hasn't understood much of the last few days of lecture, and has 85 pages of A&P to read for Wednesday's exam. She feels depressed, paralyzed by the workload and how little time there is to do it, and worried about what her parents are going to say when the grades come out and she needs tuition for the next semester. She's starting to fear that she'll never bring her GPA up and won't get into nursing school. Overcome by guilt and low self-esteem, overwhelmed by the sense of how hopeless her situation feels, Janet opens a bag of cookies and logs on to the Internet to find friends to commiserate with.

Like many students, Janet takes no control over her time. Other than her class meeting times, she has no schedule, she cannot resist distractions, she doesn't allow time to exercise or even eat right, and she allows large amounts of time to slip away unproductively. She gets caught in a downward spiral of falling behind, feeling guilty, not tackling her work because it seems hopeless to even try to meet her deadlines anyway, thus falling further behind and feeling still less self-esteem.

A Well-Scheduled Day: Let's contrast this with another hypothetical student, Joel, who has learned to value his time:

7:00 AM Joel has slept eight hours and wakes up spontaneously a couple of minutes before the alarm clock was to go off. He shaves, dresses, gets his books together, and goes to the cafeteria for breakfast. Over breakfast, he reads a newspaper (including the editorial page—not just the comics and sports sections). He gets to his 8:00 history class five minutes early and spends the time going over yesterday's lecture notes, underlining a few key terms, until the professor arrives.

8:50 AM Joel has an hour's break before calculus. He goes to the library to a quiet carrel way in the back of the third-floor research stacks. He spends five minutes going over that day's history lecture notes, checking key dates and names and making note of a question that he wants to ask in tomorrow's lecture, then works calculus problems before going off to his 10:00 calculus class.

10:50 AM Joel goes back to the dorm, changes, and goes jogging for half an hour around campus. He returns to the room at 11:30, showers, and reads half a chapter of history.

12:30 PM Joel joins a couple of guys from down the hall for lunch.

1:15 PM Joel goes back to the room. He has a 2:00 anatomy lab on the skull, so he previews that chapter of the lab manual and textbook, getting a preliminary idea of the bones he'll have to locate in lab, making a list of the major ones, and briefly quizzing himself to see if he can visualize where the bones are. For the last half-hour before class, he reads from the textbook on calcium metabolism and ossification of the bones.

3:50 PM Joel and his study partner go back to the cafeteria, which is relatively empty between serving hours. They get some coffee, quiz each other on cranial anatomy by covering up textbook labels and asking each other to name the bones, make a short list of the bones they didn't know very well and need to study again in lab, and spend a little time reading physiology.

5:00 PM Joel goes back to his room and takes a nap for half an hour. His roommate comes in at 5:30. They shoot the breeze, watch the local and national news on TV until 6:30, and go down for dinner.

7:30 PM Back in the room, Joel watches a favorite half-hour sitcom on TV, then turns it off. His roommate goes out for an evening class and Joel spends an hour reading A&P and taking notes from the book. His roommate and a friend come in at 9:00, and they're a bit noisy. He talks with them for a little while, but at 9:30 Joel takes his history book and goes back to the library for an hour.

10:30 PM Joel returns to the room, explores a few music websites, and goes to bed at 11.

In Joel, we see a student who doesn't waste time, yet he's not an obsessive "workaholic" who has no time for friends or recreation. He's spent an hour and a half watching television, half an hour on the Internet, and half an hour on physical exercise; worked in a short nap; and spent almost four and a half hours with friends (some of that combined with meals and study). All in all, he's had a pretty full extracurricular day, but he also spent four hours in the classroom (two lectures and a lab) and spent **six and a half hours studying outside of class** (in contrast to Janet's hour!). Notice that Joel doesn't waste small amounts of time. Even a few minutes at breakfast or waiting for an instructor to arrive for class are spent usefully.

A person who has this much balance in his or her life and this much control over the daily routine isn't likely to let exam dates sneak up unnoticed the way Janet did. He or she is likely to be more confident, better prepared for major exams, and on track with assignments, such as term papers. This kind of person is more likely to earn good grades and feel better about himself or herself than one who lets a lot of time slip by with nothing to show for it.

Budgeting Your Time: Presumably, you know what bills are coming up, what it's going to cost you to eat and buy gas for the rest of the week, and so on. You don't spend your money on every little whim that comes along without regard for what the necessities cost. You shouldn't spend your time that way, either. You should know what exams are coming up and how much time it will take to be ready for them.

How are you going to spend your allotted 24 hours each day? Nobody expects you to have every hour tightly scheduled, or to devote every waking hour to school at the cost of leisure or friends. But you should have a clear idea of where most of your time is going. Let's break the day down into broad categories of activity and make some rough assumptions about the time they require.

8 hrs	Sleep
2 hrs	Meals; half an hour each for breakfast and lunch and an hour for dinner
2 hrs	Self-maintenance; bathing, dressing, laundry, cleaning up, etc.
3 hrs	Class
5 hrs	Studying
2 hrs	Socializing
2 hrs	Leisure, recreation, exercise and open time; reading, sports, television, games, etc.

A good rule of thumb for planning your study time is to allot **two hours of study for every hour you spend in class.** Most labs probably do not require that much outside study time; for example, it should not take four hours to write up a report on a two-hour physics lab. However, anatomy labs may take more outside study time than many other lab courses. The five-hour estimate bears in mind that you will also study on the weekend. Often, you can increase your efficiency by combining activities—for example, you can socialize at meals or during recreation, read textbooks while waiting for your laundry, review vocabulary flashcards while eating lunch, and so on.

Note that this schedule still leaves open 32 waking hours during the weekend. This can be used to accommodate family, other social or community activities, extra study or research for term papers and exams, and so forth. If you have more responsibilities than a dormitory resident—for example, you commute long distances to campus, you have to cook your own meals and wash dishes, or you have a spouse and children—there is latitude in this schedule to accommodate them. If you have a part-time job, you should realize it will require compromises in the social and leisure time, though it should not require that you give those up entirely. If your job takes more than four hours per day, however, you probably need to consider taking less than a full course load. College doesn't *have* to be completed in four years. Notice that the in-class and study time add up to eight hours/day Monday–Friday, or 40 hours a week—on a full course schedule; **college *is* a full-time job.**

The key elements in getting control of your time are (1) planning ahead, not waiting until you have the time to do something but *making* the time to do it; (2) sticking to your schedule; (3) resisting distractions; and (4) not letting small periods of time go to waste, mistakenly thinking there's not enough time to do anything. Bear in mind the following pointers when you're deciding how to apportion your time.

1. **Plan for recreation, exercise, and leisure.** These shouldn't be overlooked or left to chance but should be part of your routine. You should be able to anticipate when you are going to do them. Don't neglect exercise, even if it is as simple as walking! Physical exertion relieves stress and increases overall well-being.

2. **Think about the best time of day for the tasks you must attend to outside of class.** Some people concentrate best in the early morning and others late at night. Plan your study time around these hours and save "no-brainer" activities—shopping, cleaning, e-mailing, etc.—for hours when you're less intellectually efficient.

3. **Don't study for too long at a time or study one subject for too long.** We become less efficient when we overdo it. It is best to switch to a different subject after an hour or so, preferably to a subject quite different from what you have been studying. For example, after an hour of A&P, take a 10-minute break and then read an English assignment for an hour, rather than studying A&P for two consecutive hours. After two hours of combined study, consider breaking for a nonstudy activity that gets you up and moving around—perhaps household chores or a walk. Often, you will feel much better and more anxious to hit the books again after a swim, a movie, or dinner with friends.

4. **Don't waste little bits of time.** Successful students find ways to use all their time wisely. Some go through flashcards while waiting in the checkout line at the grocery store or listen to lectures they've recorded while commuting or exercising. It may not seem that it's worth trying to study if you only have 10 or 15 minutes, but it's surprising what can be accomplished in little bits of time like this, and if you utilize them, they add up to a lot of hours over the course of a week.

5. **Don't allow other people to waste your time by dropping in to chat or keeping you tied up on the phone.** A tactful "Can we talk later?" will avoid alienating friends while it puts you back in control of your time.

6. **Keep a wall calendar by your desk—or any calendar on which you can see a month at a time.** Mark this with major due dates and exam dates, and look at it daily, so that these things won't sneak up on you.

7. **Get a weekly planner**, carry it with you to classes, and mark due dates, appointments, and so forth in it as you become aware of them. Each academic term, make up a weekly schedule. Fill in your class schedule first; then block out times that you're going to set aside for study, meals, recreation, employment, and the other basic tasks discussed earlier. Keep it in plain view at your desk and *follow it*.

As you develop a routine, learn to screen out distractions, and say no to the time-wasters in your life, you will increasingly feel the rewards of well-focused work. Your mind will wander less from the work before you, you'll get a sense of gratification from how much you get done, you'll sleep better not feeling guilty or worried about what you didn't do, you'll feel less pressure in the days just before an exam, and you'll gain self-confidence.

Nontraditional Students: Some of what has been written above applies especially to students 18–22 years old who begin college immediately after high school. However, many A&P students are "nontraditional" students, often beginning or returning to college after career or family experiences and often with spouses and children. These students are often among the most mature, motivated, goal-oriented, and focused members of a class, all which work in their favor. On the other hand, they also tend to be more self-supporting and have heavier demands on their time from home, family, and jobs. Younger students sometimes attend college because someone else expects it of them. Nontraditional students are almost always doing it because they want to—because they aspire to get more out of life than the alternatives they've already experienced.

Some nontraditional students are needlessly apprehensive about college, while some are overconfident. Those in the first group may fear that they can't compete with younger, more agile minds and won't be able to keep up with them. This isn't true. Our joints may stiffen, but our minds get better with age and experience. Some may fear that their instructors will expect more of them, just as we may expect more of our 16-year-old child than we do of our 10-year-old. This, too, is untrue. You will probably find that your instructors set the same standards for everyone; they don't raise the bar higher for older students.

If you are a nontraditional student, you may have more experience and a broader perspective than your younger classmates. You are probably closer to your instructor in age, perhaps even older than some of your instructors, and you may view them more objectively than younger students. However, don't assume that they will be more impressed with you because you're older, and don't assume you know it all and just need a certificate to document it. Most instructors welcome a mixture of ages in their classes and often find that older students have a stabilizing influence. However, even people already certified in some health care professions, such as emergency medical technicians, do not already know it all or necessarily earn the highest grades. Whatever your age and experience, you still have a lot to learn and you will still be challenged by A&P. Don't underestimate the time and discipline it will require or the relevance of the foregoing advice to you.

Students with families, especially with dependent children, face particular challenges. It's easy for parents to feel overwhelmed by the competing demands of school and family and guilty about the time taken from their spouse or children. Perhaps the solution to this, as for so many other things, is free and open communication about your educational goal and experiences. Make your family a part of the process; involve them in your aspirations, help them see the value in working hard for a worthwhile goal, and take pride in being a role model for your children. Among other things worth discussing in the family are the following issues: Do they support what you're doing? Do they see rewards for themselves in your goals? How do they feel about the time you take away from them for classes and study? How do *you* feel about that? Can you schedule activities and ways to be together regularly? Can they help with the responsibilities for the household? Perhaps your spouse or older children can help test you when you're preparing for exams—for example, by reading passages from your flashcards and seeing if you can supply the right terms. Even if you don't necessarily need that help, by involving your family you can make it feel like a group adventure. You'll feel less stress over shutting them out while you study and this will translate to feeling more comfortable with the time you spend in the classroom or hunched over the books.

Internet Addiction: Generations of students have gone to academic ruin spending long hours playing cards or video games. Today, the fatal attraction for many seems to be the Internet. It can be a valuable research and learning tool or a destructive seduction. We certainly didn't have an anti-Internet attitude in developing this book. We have established a home page for it with links to a multitude of websites for further information related to each chapter. Many students provided valuable insights and suggestions for Internet use while it was being developed. Most communication between the textbook author and editors was handled by e-mail, yet they also encountered many depressed and failing A&P students on the Internet. It was clear that their obsession with the Internet was pathological, much like compulsive gambling or channel surfing. There was an obvious difference between the successful students and the failing ones. The successful students were logged on only occasionally and for limited periods. The failing students were logged on frequently and for long periods, and they often admitted to their own "addiction" to the Internet and feeling of helplessness over it. Clearly, for many college students, easy access to the Internet has done more harm than good.

The Internet has spawned its own prolific humor: you know you're addicted to the Net if all of your friends have an @ sign in their names; you never write to your mother because she doesn't have a modem; your cat has its own web page. But for many, it is no laughing matter. You may have Internet Addiction Syndrome if (1) you feel habitually and irresistibly drawn to the Net; (2) you repeatedly pledge to cut back your Net time and repeatedly fail; (3) you feel gloomy when you don't log on or can't find any of your usual friends logged on; (4) you're spending hours and hours on the Net to the detriment of your grades, work, or social relationships; or (5) your relationships with people on the Net feel more meaningful than your relationships with people you see every day in "real life."

The Internet should be a resource, not a lifestyle. You must decide what you want to be doing two, three, or five years from now and whether your Net time lessens your chances of achieving that goal. If you have a self-destructive preoccupation with the Net, there are websites that can help you get it under control. This may sound like giving an alcoholic directions to the liquor store. (As the joke goes, "Addicted to the Net? Find help at http://net.addiction.com!") However, the following sites have information that will enable you to recognize the signs of addiction and to get in touch with support groups of other people working to overcome the same habit.

http://www.netaddiction.com
The web page of the Center for Online Addiction (COLA) is directed by Dr. Kimberly S. Young, an expert on Internet addiction. Her web page includes links to a survey you can take to evaluate your problem and a link to a live "chat room" for an Internet addiction support group, among other sources of information and help.

http://psychcentral.com/netaddiction/
This site offers more information and links to other sites on Internet addiction.

Although they can provide you with further information and insight into the problem and put you in contact with other people striving to deal with it, any real help must be found among groups *away* from the Internet. Check with your **campus counseling service** for referral to on-campus or off-campus support groups and recovery programs. Most communities have such groups. Their members may be struggling with a variety of issues ranging from alcoholism or drug use to compulsive gambling or shopping addiction. Nevertheless, there are many common threads between these and Internet addiction, and such groups can offer effective strategies for dealing with the issue.

Getting Off on the Right Foot

There are some things you can do in the first day or two of class, or even before class has met, to set the tone for a successful term.

Your Textbook: When we begin a journey, we want some impression of the destination. The same should be true of beginning a course. When you get your textbook, thoroughly inspect it, so that you'll know what it has to offer and how to get around between the covers. Take note of the glossary, appendices, and reference materials printed inside the front and back covers, which you may find useful very early in the term. Read the Preface to Students, where some pointers are given on making the most effective use of the book. In the Preface to Instructors, there is a list of supplemental materials available to aid in your study. You may wish to order one or more of them early in the term if they're not already available in your campus bookstore. Note the content and sequence of the chapters, and if class has met and you have a syllabus, make a note of how much of the book your course will cover.

Ideally, **buy a copy of the book new and keep it** at the end of the term. You don't want to be distracted by the markings of a previous owner, which are sometimes quite extensive and annoying in used books.

For professional reference books, the savings realized on a used book probably don't outweigh the benefits of having a clean copy and making it uniquely yours. If you go on to nursing or another professional school, you will find that instructors expect you to know your basic A&P; they do not have the time to reteach it. You will be taught about clinical aspects of hypertension on the assumption that you already know about renin, angiotensin, and aldosterone. Unless you have a functioning photographic memory, you'll find yourself needing a book where you can quickly refresh your knowledge. Your A&P textbook should become a permanent part of your professional library.

If at all possible, read the first 10 pages or more of the textbook before the first day of class. This will allow you to be ahead of most of the class, and it will give you a feeling of confidence to recognize some of the terms and concepts on the first day of lecture. Getting off to a good start like this can help create the positive attitude that will energize and motivate you for the rest of the term. While the rest of the class is scrambling to read pages 1–10, you'll already be reading beyond that.

Your Notebook: For class notes, the choices are spiral-bound notebooks or three-hole loose-leaf notebooks; some students use laptops for in-class note taking. With spiral-bound books, it is easy to run out of paper (there is a lot of material in an A&P lecture/lab course!); it is also difficult to keep and organize any handouts given in class. Loose-leaf notebooks allow for easy expansion and the addition of any class or lab handouts; they also let you add pages when you review for exams. The disadvantage of laptop note taking is that you can't draw diagrams, an essential part of most A&P lectures. **Develop a system that works for you.** Whatever system you use, stay organized; keep your notes in order and ensure that none will be lost. Keep your A&P notes in a notebook separate from your other classes. Put at least your name and perhaps your phone number in your notebook (and text). Unidentified notes/textbooks left behind in a classroom cannot be returned. You have a large investment in this course; protect it!

Your Schedule: When you meet the instructor and get a course syllabus, immediately mark all known exam dates and assignment due dates in your weekly planner and on your wall calendar. Don't let assignments and exams sneak up on you.

Your Study Partners: Within the first few days of class, you may want to gravitate toward one or more people with whom you would feel comfortable studying. Study partners can be a help or a hindrance, depending on how well you choose each other and how effectively you use your time together. Choosing the right laboratory and study partners can make a significant contribution to your success, especially if you structure your joint study time well and establish a system of testing each other on textbook and laboratory material. Two minds are better than one—you can exchange interpretations, fill in gaps in each other's comprehension and knowledge, and divide the labor of preparing review lists or other materials.

It is important, however, to choose a partner carefully, as the following quotes from students will demonstrate. If possible, team up with one or more people who have a study style compatible with yours. If you are serious about getting as much work done during a study period as possible, choose partners who will stay focused on the subject and not waste time on irrelevant chatter. **Don't be afraid to change partners or groups** if necessary.

> I used to study with Sherry and Ann, but they'd spend half the period running their mouths about everything *but* anatomy. It was a big waste of time.

Choose a partner whose level of ability is similar to yours. It may be tempting to buddy up with the people who got high scores on the first exam, but if you're struggling to get a *C,* you may find them moving along so fast you cannot get much benefit out of your study sessions with them.

> I tried studying with Phil and Nancy, but they were just so far over my head. They'd move on so fast I couldn't follow them.

You may benefit more from studying with someone who is having the same problems you are, because you will both understand that you need to work at a similar pace and drill yourselves over the material more than the faster students do. However, if you're one of the fast learners, don't let yourself be held back by someone who is a great deal slower.

> Jeff is my study partner, but I feel like I'm doing all the work. It's frustrating to me because I need to move on to new material.

Stay focused when you study together. A study session is not a social hour or gossip-fest. Agree on the rules early and follow them. It is probably best not to have more than three or four people in a group. With more in a group, almost invariably at least one person either feels excluded from the discussions or tends to coast and try to absorb ideas from the others without contributing much. If your school has a tutoring service, take advantage of it. Most tutors will work with small groups. **Don't study with a boyfriend or girlfriend.** It's hard to focus on the task if you're in love with your study partner.

Getting the Most Out of Lectures

One of the instructor's principal functions is to present information orally in class or moderate discussions that promote learning. While there are various ways to do this, lectures are still the most common approach. A skillful instructor will adjust lectures to the vocabulary, level of comprehension, note-taking speed, and questions of the audience. But a **successful lecture also requires a skillful listener**, and here we will consider some ways of becoming one.

Lectures and exams need not be your only contact with the instructor. Most institutions require that the faculty have office hours, during which they are receptive to student visits. Don't take up an instructor's time needlessly for a daily chat or to go routinely over things you understood, but if there is something in the lecture or textbook that you *didn't* understand, by all means ask for clarification. Find out early (the first day of class) what is expected of you and what you can expect of the instructor. This information is usually in the syllabus. Read it carefully and ask questions if you don't understand something.

Also take advantage of other special aids the instructor offers—practice questions, small-group reviews, question-and-answer sessions, and so on. Most instructors have noticed, with considerable regret, that when extra review and Q&A sessions are offered, students who really need the help seldom attend. This makes it harder for instructors to be sympathetic to their difficulties.

Organizing Your Notes: In counseling academically endangered students, instructors often see that their lecture notes are very disorganized. Different courses are haphazardly mixed together, with A&P notes immediately followed by history or algebra notes; notes are undated; it's hard to tell where one lecture ended and another began; they sometimes have to rummage through a notebook or backpack, looking for the notes they took on a certain day or topic; and so on.

When you are taking notes in lecture or from the book, **start a fresh page for each day**, date your entry, indicate whether it consists of book or lecture notes, and give the entry a short title—for example,

| 3/12/09 | Cardiac Cycle | Lecture |
| 3/26/09 | Anatomy of Kidney | Text |

You may want to wait until the end of a lecture to title your notes, since it's hard to predict in advance how far that lecture will go and what it will cover.

Taking Lecture Notes: The following pointers will help you keep up with the lecturer and produce a set of notes that will be meaningful and valuable for your later review. These are suggestions; **develop a system that works for YOU.**

1. *Go prepared.* If you know the topic of the next day's lecture, at least preview that part of the book before you go (and all the better if you can thoroughly study it, thus keeping your textbook study a jump ahead of the lecture coverage). **A lecture will mean a great deal more to you if you're not getting the information there for the first time** but, rather, it covers the high points of something you've already read about. Your lecture experience will be more enjoyable and less frantic because you'll feel less compelled to try to write down everything that is said if you already know that it is clearly explained in the book. If you haven't read the book, you'll more likely struggle to take down every word of lecture because you won't know whether this is your only chance to get the information. Unprepared students suffer writer's cramp, while the better-prepared student may need to write only "Explained Golgi tendon reflex, see pp. 506–507."

2. *Format your notebook pages.* Some students find it effective to keep their lecture and textbook notes in parallel. One way is to take lecture notes only on the right-hand pages of the notebook, then look up the related textbook material and write those notes on the left-hand pages, alongside the lecture notes on the same topic. A variation on this is to draw a line down each notebook page two or three inches from the left side. Take all your lecture notes to the right of the line; then when you study the text, make notes, annotations, text references, and so forth on the left, alongside the related lecture material. There are possible drawbacks to this method. For one, it implies that you will generally be hearing a lecture before you take notes on that subject from the book, whereas we have already seen there are benefits to covering the book before lecture. Another is that your lecture notes may not leave enough vertical space for all that you consider relevant to write down from the book. Nevertheless, the parallel notes technique is very successful for some students. When you later annotate your lecture notes from the textbook, be sure to write down page numbers. This makes it easier to go back and find details later.

3. *Take your book to lecture.* Keep the book open to whatever the instructor is lecturing about and follow along in the book as much as possible. Mark the book margins with important points made in lecture. In your lecture notes, make references to the book—for example, "fig. on 530." You won't get bogged down in lecture as much if you can make reference to the book rather than having to copy every sketch or flowchart from the board. Some students successfully use Post-it™ notes to mark important ideas in the text during class, making them more permanent later.

4. *Write in ink.* Inked notes are usually easier to read and look neater. Use a color that will be easy to read when you're tired and studying late at night. (Save your pink and purple pens for doodling.) There's nothing wrong with crossing out errors made in ink, and it's faster than flipping a pencil over to erase during a lecture. Furthermore, if your notes are in ink, when you study later you can annotate them in pencil, which can then be erased without erasing the original lecture notes. For example, you might want to put pencil checks by important points you need to drill on

and later erase those as you feel you've mastered that material, narrowing down your list of "things I don't know yet" as the exam approaches. Write only on one side of the paper. (See item 2.)

5.   ***Develop your own shorthand.*** Don't try to write lecture notes in a narrative form; you won't be able to keep up very well, and you'll miss information. Develop and use your own personal shorthand of symbols and abbreviations. Some common symbols you can adopt from mathematics, chemistry, pharmacy, stenography, and so forth include @ for "leads to"; ↑for "increases"; ↓ for "decreases"; Δ for "change in"; b/c for "because"; w/ for "with"; w/o for "without"; n. for "neuron"; m. for "muscle"; a. for "artery"; v. for "vein"; and so on.

6.   ***Outline.*** Don't write lecture notes in run-on paragraphs as if to paraphrase what the lecturer said. Use outlining, indenting, line spacing, and so forth in a way that makes the structure of the topic easy to review so you can see how facts relate to each other. If you have advanced knowledge of what the book covers, your lecture notes can often take the form of simple reminders of what was discussed and what you need to study. It may be enough to write "Na^+–K^+ pump—Study!" especially if you were following in the book and you know that what was stressed or discussed at great length in lecture is fully covered there.

7.   ***If it's on the board, it's probably important.*** When an instructor writes something on the board, it is a sign that he or she regards it as a major point, term, diagram, or concept. Write it down in your notes. Instructors are most likely to test on things they write on the board, not necessarily because they consciously remember to test what was on the board but because in both situations—giving a lecture and making up an exam—the same things tend to stand out in their minds as having major importance. The same impulse that prompts a lecturer to put something on the board is likely to guide his or her hand when writing the exam.

8.   ***Don't make assumptions.*** The fact that the instructor didn't lecture on something doesn't necessarily mean it's not important and won't be on the test. Lecture time is precious. If something is clearly explained in a textbook, an instructor may not feel the need to explain it in lecture but will use that time to cover the more difficult concepts. Thus, you can probably expect lecture to spend a lot more time on the mechanisms of enzymatic digestion than on the anatomy of the stomach and intestines, yet find that the exam covers a lot of anatomy.

9.   ***Ask questions.*** If your class size and format allow for questions during lecture or at the start of the period, don't let the opportunity go to waste. **Students who do best on exams tend to be those who ask questions in class.** Don't be afraid that anyone will think you're stupid; other students will probably be grateful that somebody asked. Instructors appreciate students who speak up in class, seem involved in the material, and provide relief from a sea of blank, silent faces. However, don't go to the other extreme and ask needless questions just to hear your own voice or call attention to yourself. The instructor's pleasure in having students speak up now and then may give way to annoyance if he or she is excessively deterred from getting through the day's agenda.

10.   ***Review your notes immediately.*** Most forgetting occurs in the first 24 hours after a lecture. If you can **take a few minutes at the end of class to go over your notes**, clarify any ambiguous symbols or scribbles you wrote, and refresh your memory of the main things that were covered, you will remember the material much better later on. When reviewing, put a *Q* or some other notation (preferably in pencil) beside points that you didn't understand. Try to answer these questions for yourself by reading the book later that day, but if you cannot find or understand the answer, ask the instructor on the next class day before he or she goes on to new material. Most good instructors will be pleased that someone remembered the point and cared enough to ask about it. Even if you can't review the notes immediately after class, do it sometime later that day, before accumulating a new day's lecture notes on top of the old.

11. ***Should you recopy notes?*** Some experts on collegiate learning advise that you not spend time recopying (much less typing out) lecture notes. This can eat up a lot of your time without a proportionate benefit in learning or grades. On the other hand, some of the best students routinely do this because their in-class notes are relatively sloppy, and this provides a neater set to study from. Try it if you think it will help, but don't get fixated on this strategy and don't be afraid to give it up if it takes more of your time than it proves to be worth.

12. ***Should you record lectures?*** Policies on recording lectures vary. Some instructors prohibit it, while some institutions have policies that guarantee students the right to record lectures. Check your instructor's policies. Recorded lectures can be a valuable way to fill in gaps in your notes, but beware: Some students actually hurt themselves with recorded lectures because they feel compelled to play them back and write down everything the instructor said. This means they often stop and "rewind", making sure they get everything down and thus spend two to three hours transcribing what the instructor said in 50 minutes. This can be a devastating drain on your valuable time. **Don't use recordings as a substitute for taking handwritten notes in lecture**, with the idea that you'll copy everything down later. If much of your day is spent commuting, perhaps the most valuable use for recorded lectures is listening to them during the drive. *Do not,* however, try to look at notes or jot things down while driving.

Getting the Most Out of Your Textbook

Anatomy and Physiology: The Unity of Form and Function is designed not just to provide you with information but also to help you in the learning process. Many features are built into each chapter for this purpose, and the objective here is to discuss these and how you can use the book most effectively.

Above all, don't expect to do well in an A&P course without reading the textbook. Instructors are repeatedly surprised by students who refuse to believe this until they have already failed an exam or two. Students often ask, "Do you think it would help if I read the book?" Perhaps they found in high school that they could get by without reading their textbooks much, but it doesn't work in A&P. Lecture time is very limited; it isn't nearly enough to adequately cover a subject as complex as this or to cover everything a student needs to know before pursuing other health science courses. You should **view lectures mainly as an opportunity to hear textbook concepts expressed in a different way**, to get some clues as to what your instructor thinks is most important, and to ask questions about things that give you difficulty.

Now let's examine the structure and features of the textbook chapters and how to integrate them into an effective study plan. Chapter 18 will be used as an example in much of the following description.

Chapter Outline: The outline on the first page of each chapter will give you a sense of your learning goals as well as a plan of attack. You'll note that each chapter is divided into about five or six sections. These average five to six pages each, although a few are significantly longer (especially if they entail anatomical reference tables) and some are as short as one page. Think about when you must have the chapter finished, how long the instructor is likely to spend on it before moving on to the next chapter, and which sections you will read each day. The outline will also enable you to break the chapter down into reading segments that correspond to the time you have available. If you know the instructor is likely to spend only two or three days lecturing about blood, for example, then you had better plan to read about plasma and erythrocytes one day and leukocytes, platelets, and clotting the next day, thus completing the chapter by the time he or she moves on to cardiac physiology.

Reportedly, the average person reads technical material at about 100–150 words per minute (wpm), compared with 250–300 wpm reading a novel, newspaper, or other nontechnical material. *Anatomy and Physiology* averages about 50 words per column inch, not counting illustrations, tables, or special features. The section on erythrocyte disorders (pp. 689-690), for example, is about two pages of text—approximately

1,370 words. An average reader should be able to read this, if not taking notes, in roughly 15 minutes. For a person taking notes, the time would be highly variable depending on writing speed and amount of detail written down. If you pay attention to your own speed, however, you will soon develop a feel for how long it will take you to read five or six pages with note taking. That self-knowledge and the chapter outlines should enable you to plan your study time fairly accurately. Don't forget to study the diagrams and tables.

Brushing Up: Each chapter from 3 to 29 has a feature titled "Brushing Up" at the beginning. Each chapter builds on knowledge imparted in earlier ones, so if you are not conversant in those topics, you might have difficulty with the new chapter. On page 679, for example, you can see that the study of blood requires a prior knowledge of polypeptides, conjugated proteins, osmosis, filtration, dominant and recessive alleles, and sex-linked inheritance. If your memory of those topics is vague, or you haven't covered some of them yet, then it would be advisable to go back and review those before continuing.

Expected Learning Outcomes: Each section of the chapter begins with a list of objectives or expected learning outcomes (ELO). Read these so you will know the major topics that a section is meant to convey. After studying a section, go back to them and see if you have met those goals.

As You Read: You **should not have to read a text more than one time IF** you are prepared properly. There are several reading/study methods available: SQ3R (Robinson, F. P., *Effective Study*, 4th ed. New York: Harper & Row, 1970), PQRST, and others. All of these methods take some effort on your part but are well worth the time spent. For example, using the PQRST method, begin an assignment by previewing (P) a section in the text: Read the headings and expected learning outcomes, looking at figures and perhaps their legends. While you do this, ask yourself questions (Q) about what the headings mean or, since the text has ELOs already written, think about what the section is going to reveal about them. Then **read with a purpose** (R). The purpose is to answer the questions you asked or be able to explain/answer the ELOs or "Before You Go On" questions. For example, on page 684: Erythrocytes, you'll notice that the first ELO asks you to describe the structure and function of RBCs. The answer is clearly outlined on pages 684-685. After you read a section, state (S) or, better yet, **write out the answers to the questions** you asked yourself. Writing out the answers rather than highlighting the text is an ACTIVE method of learning. The (T) in PQRST is for *test.* When you finish doing P, Q, R, and S, check yourself to see how much you remember about what you have just read and written. Page-reference your notes, and make note of related figures. This will help you later; you can quickly go back to where you read something and fill in details that may not be in your notes.

These reading/study methods force you to be an active participant. They seem like time-consuming ways of reading a text but, with practice, **you will actually save time.** Your college learning or counseling center will have more resources and can help you learn how to improve your reading/study skills. Take advantage of these services!

Marking the Book: Books are not sacred. Perhaps in high school, if the school system supplied the books and you were expected to return them unmarked, you were conditioned not to write in your books. But in college, you buy the book—it's yours, and you can add to its value by making it personal. Make marginal notes that will help focus your later review and study. Put asterisks by important material. Write a *Q* or *?* in the margin by things that give you difficulty or that you want to ask about in class.

Many students highlight their texts with colored markers. **Highlighting is passive** and will be of little benefit to the majority of students. In time, highlighter ink becomes darker and makes text difficult to read. Some students highlight so much of the text that it's just as useless as highlighting nothing—nothing stands out.

Vocabulary Aids: Your exam questions will include technical terms; if you don't understand them, you won't understand the question, much less the answers. Memorizing words and definitions is not the same as really understanding A&P or any other discipline. Nevertheless, to be conversant in any field, you must have a working knowledge of its vocabulary. Therefore, the textbook has several features designed to help you master vocabulary more efficiently and become more comfortable with the language of A&P.

Boldfaced and italicized terms. The most significant terms in each chapter are set in **bold** print. Pay special attention to these, and when you complete a section, ask yourself if you can define each of them from memory. Make up an original sentence that would use that term, as if you were the instructor making up a sentence completion test question. Terms set in *italics* either are of secondary importance or will be discussed more extensively (and often will appear in boldface) later on. Italics generally imply "This is what ___ is called, but the author does not consider this a term that most A&P students need to memorize." However, instructors differ in how much importance they attach to terminology. If your instructor stresses it in lectures, then, of course, you should give it more consideration.

Pronunciation guides. It is difficult to remember, spell, or define words that you can't pronounce in the first place, so pronunciation guides have been provided (in parentheses) to many of the new terms introduced. These were developed with input from students to make them as easy as possible to pronounce. Simply accent the syllables in capitals and pronounce them phonetically—for example, pro-NUN-see-AY-shun. The more you sound these out, relate them to the spelling of a term, and use them in discussion with your peers, the better you will remember them on tests.

Word origins. It is much easier to remember, spell, and define words if you recognize familiar roots in them—for example, if you recognize that *hypernatremia* is composed of *hyper-* (excess) + *natr-* (sodium) + *-emia* (blood condition). Inside the back cover of the textbook is the "Lexicon of Biomedical Word Elements," which lists roots, prefixes, and suffixes. Throughout the text, you will find new terms footnoted to show their meanings and origins. Read the lexicon at the beginning of the course, study the footnotes habitually, and make it second nature to look for familiar roots in new terms. This will help you to a great extent not only in this course but also in more advanced ones. There are two types of word origin exercises in this *Student Study Guide.*

Glossary and supplemental dictionaries. Make good use of the glossary (pp. G1-G20) when you encounter terms whose meanings you have forgotten. Also keep a good English dictionary close at hand; *Merriam-Webster's Collegiate Dictionary* defines many A&P terms. A thorough medical dictionary, such as *Stedman's Medical Dictionary* or the more nursing-oriented *Taber's Cyclopedic Medical Dictionary,* is an excellent investment, available in most major bookstores.

Deeper Insights: Throughout a chapter, you will find occasional **Deeper Insight** boxes meant to add human interest, clinical relevance, and background to the core concepts. See, for example, the biographical sketch of Charles Drew on page 692 and a discussion of bone marrow and cord blood transplants on page 695. At the end of each chapter is a longer clinical application, but it serves the same purpose as the shorter ones.

Key terms included in the Deeper Insights are not in the vocabulary list at the end of the chapter or in any of the test questions in the book. However, they may be found in some questions in this *Student Study Guide*. Also, your instructor may feel that some of these topics and terms are of central importance and may test on them.

Illustrations and Tables: Pay attention to the illustrations; they're not there just for decoration. An illustration is a concise way to visually sum up a lot of information, explain an important idea, and help you see the relationships between physiological events. When the text cites a figure, be sure to study it. Much of what you read will make little sense if you don't carefully study the figures but should make perfect sense if you do. Often, the text is only a capsule summary of information that is to be gotten primarily by studying the illustration—for example, on the anatomy of a tooth (pp. 959–961). Be sure to read the figure legends carefully. Many figures show physiological processes and the accompanying descriptions in the text provide details of the processes; see pp. 412–416, Figures 11.8–11.10, for example. The combination of text and figures makes complex processes much easier to understand and makes excellent reviews before exams.

Furthermore, many people are better visual learners than verbal learners; you may find it easier during an exam to remember what something looked like than to remember the words in which it was described. This applies even to something as simple as remembering on which side of the abdomen the appendix is located. *Right* and *left* are easily forgotten, but you may find it easier to visualize the familiar illustration of the large intestine. In your mind's eye, you can see the appendix at the lower left side of the page and the sigmoid colon at the lower right, so you know that, anatomically, the appendix is on the right and the sigmoid colon on the left. As an aid to learning anatomy, try photocopying the illustrations. Cut off the labels, recopy the trimmed figures (perhaps make a few duplicates), and then see if you can label them from memory. Figure exercises included in this *Student Study Guide* challenge your knowledge of anatomy and the functions of some of the structures.

Tables such as 8.5 on p. 268 provide useful information in an easy to read format. More complex tables like those in chapter 10 are associated with figures illustrating a specific group of muscles. (See Table 10.2, p. 327.) These pack a lot of material into a small amount of space.

Apply What You Know: The true test of your understanding of A&P will be your ability to apply what you know to new situations and to think more deeply than at the level of just memorizing terms. One way to check your comprehension is to answer the "Apply What You Know" questions dispersed through each chapter (for example, p. 686). These questions prompt you to think about the deeper implications or clinical applications of something that has just been explained. They may relate to what you have just read or to something that was explained in an earlier chapter. Instructors have the answers to these in the *Instructor's Manual*.

Before You Go On: When you reach the end of a section, try to answer the "Before You Go On" questions from memory. It will be best if you write these out. (You can do this as part of your "stating" in the PQRST study method.) When we answer things "in our heads" we tend to be more fragmentary than we realize; you can better judge your ability to answer such questions by putting the words to paper. If there is one you cannot answer, make a pencil mark by it to remind yourself that the concept didn't sink in and that you need to go over the section and find the answer. Space did not allow for an answer key to these, but you can find the answers just by reading back a few pages.

If you get several of these questions wrong or cannot think of an answer, this is a warning that you probably did not comprehend that section very well. However, getting them all right does not guarantee that you thoroughly understand that section. Again, because of space limits, these questions do not cover everything important in a section. Be alert to important concepts not tested by them.

Assess Your Learning Outcomes: At the end of the chapter, you will find "Assess Your Learning Outcomes." Here you are asked to test your knowledge of a list of topics; these are grouped by section with a page reference to the text. If you study each section as you read the text you will master the material more easily than if you tried testing yourself over the whole chapter at once.

Testing Your Recall: Each chapter has a section of 10 multiple choice, 10 sentence completion questions called "Testing Your Recall." Also, there is a section titled "Building Your Medical Vocabulary" and 10 true/false questions. The answer key to these is at the back of the book in Appendix A. Write down the answers on a separate sheet of paper, and then (and only then) check the answer key. By not marking the answers in the book, you can go back and retake the same test later, closer to your exam date, without being biased by your previous answers. These questions cover topics as evenly scattered through the chapter as possible, but a mere 40 questions obviously cannot comprehensively cover the chapter. Many important concepts and terms are left untested. This is merely a quick "spot check." A good score doesn't necessarily mean you have a good comprehension of the chapter, but a bad score *does* mean you apparently have inadequate knowledge of it and need to go back and study. Also bear in mind that these questions merely test rote memory; they do not test whether you really *understand* the chapter or whether you can apply what you have learned to new situations.

Testing Your Comprehension: Unlike the "Apply What You Know" questions mentioned earlier, the "Testing Your Comprehension" questions at the end of the chapter test your ability to think beneath the surface, to analyze and apply the ideas presented in the chapter. The answers to these are at the Online Learning Center.

At the End of a Chapter: When you have finished a chapter, go back and read the "Expected Learning Outcomes" for each section. Ask yourself if you feel you have met those goals. Read the "Assess Your Learning Outcomes" questions again. Are there any concepts you've forgotten? If you suddenly find you can't remember anything that platelets do except cause blood clotting, then you had better read your notes on platelets again and remind yourself of their other functions.

Using this *Student Study Guide*: When you have completed the recall and comprehension questions in the text and feel you know the chapter well, complete the practice exam for that chapter in this *Student Study Guide*. Collectively, the textbook and study guide provide a lot of questions on each chapter— typically, about 55–60 questions in the textbook, at least 80 questions in this study guide. Clearly, there are ample opportunities for you to test yourself, or for you and your study partners to test each other, before the instructor tests you. More than anything else, this **self-testing seems to be the key to success on A&P exams.**

However, too many students use chapter review questions and this study guide not to test their memory but to test their ability to look up answers. Consequently, they overestimate their knowledge of the subject and receive surprising and disappointing scores on exams. Anybody can look up the answers. To do that in lieu of answering them from memory is to waste the opportunity given to you and cheat yourself out of a good grade.

Take the chapter exams in this study guide as if they were your instructor's exam. Write out the answers; set a time limit. DO NOT cheat (yourself) and look up the answers. If you have study partners, grade each other's exams. Give yourself a score based on the number of questions in the chapter exams. (This varies among chapters.) If your grade isn't what you expected, study the areas where you were weak and retake the exam.

Visit Our Website: If you're researching a paper for your A&P course or simply have a personal interest in further information, visit our website, the Saladin Home Page, at

www.mhhe.com/saladin6

For each chapter of the book, you will find practice quizzes, interactive learning games, chapter outlines, and other study tools.

Supplemental Study Aids: This *Student Study Guide* is only one of several supplements available to help in your studies. Others are listed in the "Preface to Instructors" in the textbook, although some items listed there are not available to students (such as lecture transparencies) or probably would be of little use if your instructor has not already ordered or assigned them (such as lab manuals). Students can order supplemental study materials by visiting www.books.mcgraw-hill.com, calling 1-800-262-4729, or contacting their campus bookstore.

Test Preparation

You've studied a chapter, and have taken a practice test, and now the date of your instructor's exam is fast approaching. Let's look at some ways to prepare for the *graded* exam—especially how to boil down the task to its manageable essence, how to master the vocabulary, and how to test your knowledge off the record before the instructor tests you.

Rereading Is Not Studying: Too many students think that studying means rereading the textbook and lecture notes. The following dialog is typical.

Student: "I spent *six hours* studying for this exam and I *still* got a D!"

Instructor: "What did you do? How did you study?"

Student: "I read the chapter three times from beginning to end, and I read all my lecture notes over again twice. I was sure I knew it, but when I was in the room taking your test, my mind was a complete blank."

There is good news and bad news. The good news is that there *are* jobs available for people with blank minds; the bad news is that you probably wouldn't like them. So how do you use your study time effectively and not draw a blank on the exam? The student in the preceding dialog suffered from a common illusion: you can read and reread notes and texts over and over until they look so familiar to you, it's like knowing what the next line is in a favorite song on the radio. However, when all your books and notes are closed, and you have nothing but your memory to rely on, you discover, to your dismay, the crucial difference between recognition and recall. (See p. 14 "As You Read" in this SSG.) We learn very little just by reading or listening to information. We all see students sitting in the hall before an exam, gazing at the book, flipping through heavily highlighted pages, without writing anything down. It's usually a safe bet that these are the people who will not do very well. It's when we **become an active participant in the learning process,** by writing, summarizing, and self-testing, that we assimilate and retain information. To study effectively does not mean merely to go back over old material. It is a process of manipulating the material, boiling it down, identifying your weak areas, and correcting them.

The Time and Place to Study: Note the date of your exam and schedule your reading to finish at least three days before then. During the last three days before the test, you should not be covering new material but organizing and reviewing notes, drilling yourself on the information, and progressively narrowing down your "don't know yet" list. If you fail to pace yourself and then try to cram in five or six hours of study the day before an exam, you will remember much less.

You need a suitable place to study. Some find they can study best in the quiet of a library, perhaps at an obscure carrel in the back of one of the higher floors, where hardly anyone goes. To others, absolute quiet is more distracting than having background noise. Television is the most insidious thief of time. Even if it's only "in the background," it will totally steal the concentration you need to study effectively. If it's in sight, it will doggedly demand that you glance up to see what has just happened; if it's out of sight, it will make you wonder. Get rid of it.

Flashcards and Checklists: Many students find it valuable to develop a set of flashcards or a vocabulary checklist as a course progresses. As you read a chapter, make a list of terms and the page numbers where you found them. Put a pencil check by the terms you think will be most difficult for you (e.g., *gastroferritin*); don't mark those that you find easy (e.g., *hematology*). If you choose the flashcard approach, get a few packs of 3x5-inch index cards. Write one term or concept on the front of each card and a definition or explanation of it on the back. To prepare for discussion questions, you will probably need something more extensive on the front, such as "Describe how renin, angiotensin, and aldosterone regulate blood pressure," and write out an answer on the back.

An alternative to the flashcard method is to prepare a checklist of terms or concepts in your notebook. Write a descriptive phrase on the left side of the page, leave a space, and write the term on the right side:

Formation of blood	hemopoiesis
Formation of red blood cells specifically	erythropoiesis
Kidney hormone that stimulates RBC production	erythropoietin

Whichever method you use, learn to narrow down the material to the concepts and terms that give you the most difficulty. If it's easy for you to remember what the humerus and the frontal bone are, don't bother to make a flashcard for these terms or put them on a terminology study list. There are two dangers in cluttering up your study materials with easy facts. One is that it wastes valuable time. There is so much to cover, you simply don't have time to make unnecessary cards just for the sake of completeness or to waste time answering the same question each time you go through a checklist. The other danger is that a card stack or checklist with a lot of easy items in it can lull you into a false sense of accomplishment. It has so much familiar material that it blinds you to the large amount that you don't yet know very well.

Testing Yourself and Boiling It Down: Flashcards and checklists can be great preparation for sentence completion, multiple choice, and matching questions. If you know vocabulary thoroughly, you should be able to navigate those sections of an exam easily. The key is to start with the long list of things you don't know and progressively narrow it down as the exam approaches. This cannot be done in the last day or two before the exam; it must be an ongoing process, beginning with the first day of class after a previous exam.

Some of the best students carry their flashcards everywhere and use them to drill during lunch, or while waiting in the laundromat. This is a great example of how you can productively use little bits of time that would otherwise go to waste, as well as gain more study time by doubling up your tasks—laundry and studying at the same time, for example.

If your instructor asks questions in a sentence completion format—for example, "The kidneys secrete a hormone named _____, which stimulates the bone marrow to produce red blood cells"—then design your studying accordingly. Suppose your card or list says, "Kidney hormone that stimulates RBC production." Preferably, *write down* your answer on another piece of paper. Don't just "hear" the answer in your imagination, because our imaginations tend to be more fragmentary than we realize, and they often trick us into thinking we know something that we only half know. Don't even settle for just speaking the answer aloud, because this doesn't test your spelling. If you write down the answer, then you can check your recall of the right term *and* your spelling of it. If you answered *renin,* you were wrong (even though that is a kidney secretion), so keep that card in your deck, or don't check that term off your list. If you answered *erythipoten,* you somewhat know the right word but not how to spell it. Keep working on it!

The first time through, you might find you got 62% of your self-test right, which is good news and bad news—it means you've already learned more than half of your list, but not enough to get more than a *D* on an exam. Now *immediately* go back to the top of your list or card deck and start over. You will remember some things you got wrong the first time through, and on your second time through, your performance may be up to *C* level or better—perhaps 76%. You're making progress. Immediately go back a third time and test yourself on cards that are still in your stack or terms that you haven't checked off your list yet. Perhaps you're now getting 93% of them right. By this time, you probably need a break, or have English homework to do, but at least you can feel a sense of satisfaction in the progress you've made.

The next day, go back and do this again. Put all your cards back together, including the ones you got right the previous day, or erase all those pencil checkmarks from your checklist. Don't be surprised if your performance has slipped back to 78% or so. We all forget things overnight, but the nice thing is that this is much better than the 62% you started with the day before. You are retaining information from day to day. Moreover, you might now find that it takes only one repetition to get back into the 90s. If you keep this up each day leading up to an exam, you should do well on vocabulary-based questions.

Note that if your instructor asks questions in a sentence completion format, like the previous example, you should not look at the list of terms (if provided) and try to define them—you should look at the definitions and try to remember the right word. Some students tend to study "backward" and don't score very well for this reason. You must ask yourself questions in a format as close as possible to the one your instructor uses on tests. An A&P exam is not a game of *Jeopardy!* in which you are given the answer and asked to state the question.

As mentioned earlier, you can test your knowledge of anatomy by covering up the labels on the illustrations or by making photocopies, cutting off the labels, and recopying the unlabeled art to make test sheets. This is a good opportunity for members of a study group to divide up the labor, so that each person gets a complete set of anatomical test sheets but each person has to do only a small amount of the work in preparing them.

Practicing for Essay Exams: For practice in writing short essays, write out the answers to the "Apply What You Know," "Assess Your Learning Outcomes," and "Testing Your Comprehension" questions in the textbook. Keep in mind, however, that under test conditions you must be able to think and write more quickly than (yet just as clearly as) you do on practice questions. See the following discussion on test taking for further advice on writing effective essay answers.

Testing Each Other: A good study partner is a valuable asset. You can test each other with flashcards and checklists and often help each other recognize weak spots that you might not see on your own. You can also make up new questions for each other. Also, when you make up questions for yourself, you tend to be too easy on yourself. A study partner can be more demanding and persistent in discovering and going back over your weak points.

Knowing a vocabulary list is not the same thing as understanding the subject, but it is necessary. You must be able to do more than define terms; you must understand the implications and concepts behind the words. For example, one could easily memorize a definition of *homeostasis*—a state of relative physiological constancy in the body—but, to understand and be able to discuss homeostasis, you would also need to understand the meanings of dynamic equilibrium, set point, and negative feedback. The word is merely a "hook" on which you can hang a broader concept. Another advantage of studying with a partner is that you can probe beyond the surface meanings of words and demand that the other person be able to discuss underlying and related concepts such as these.

Test Taking

The day has arrived. You've kept to a schedule, you've studied the material, and now it's time to take the test. Let's consider some strategies for doing well.

First, sleep and eat well. If you've paced yourself during the weeks leading to the test, you shouldn't have to pull an all-nighter before the exam. If you get a good night's sleep, you don't have to drag yourself out of bed before your body is ready to wake up, you eat a good breakfast, and you don't have to rush to class, you'll go into the exam with a clearer head and a lot less anxiety. If possible, get to the test site early. A half-hour to go back over your flashcards or checklists just before the exam will be very helpful. If you have classes back-to-back and cannot do this, then try to set aside half an hour or so before the class(es) preceding A&P.

If there are no essay questions on the exam, answer the questions you feel confident about first. You'll feel better as you tackle the more difficult ones later. If there is a written section on your test, you may want to tackle it first. When time is running short at the end of the period, it's a lot easier to quickly answer a series of multiple choice questions than to think clearly and finish writing an essay under pressure.

If a wrong answer counts off no more than a question you don't answer at all, and if time is running out, at least guess on multiple choice, true/false, or other such questions. Don't leave answer spaces blank. On multiple choice and matching questions, even if you're guessing, try to narrow down the options by eliminating the obviously wrong answers and guessing among those that are plausible.

Answering Short Answer Questions: Make your answers legible! Some instructors are more likely to take off points than to struggle over difficult handwriting. Pay attention to the phrasing of a question and put your answer in the proper singular or plural, noun or adjectival form to complete the statement.

Students invariably ask, "Do you count off for spelling?" Merely asking this question indicates intellectual immaturity. A professional person should be able to accurately spell the terms in his or her discipline; it shouldn't even occur to you that it may be unimportant to spell things correctly. Whether you are writing on a patient's medical chart, writing a business letter, or writing a technical report, poor spelling creates an unprofessional appearance. Those who have served on hiring committees have seen people passed over for job interviews and offers because of poor spelling. It should be a habit you cultivate, not something you hope people won't count against you. **Good spelling is a mark of an**

educated person and identifies a person who will be able to communicate well and who cares enough about his or her work to pay attention to details.

Furthermore, even slight spelling errors can sometimes create significant changes in meaning. For example, the *ilium* is a bone of the hip, whereas the *ileum* is part of the small intestine; *renin* is an enzyme of the kidney, but *rennin* is a milk-curdling secretion of the stomach; and *sucrose* is a sugar, while *sucrase* is the enzyme that digests it. If you write *occipital* when you mean *occipitalis*, or *zygomatic* when you mean *zygomaticus*, you've named a bone when you mean to name a muscle. Thus, you can see what a big difference seemingly small errors in spelling can make.

Answering Multiple Choice Questions: In its simplest form, a multiple choice question consists of a statement followed by one correct response and three or four incorrect ones called foils. An important technique for answering multiple choice questions is to quickly eliminate the foils you know are wrong and focus on the answers you think might be right—for example,

Which of the following bones is found in the ankle? (a) clavicle; (b) hyoid; (c) trapezium; (d) cuboid; (e) patella

You may easily remember that the clavicle is the collarbone, the hyoid is in the neck, and the patella is your kneecap. That narrows it down to either the trapezium or the cuboid. Even if you have no idea which one is in the ankle and which is in the wrist, you can at least guess. By eliminating the others, you've improved your chance of getting it right from 20% to 50%. When you eliminate answers you know to be wrong, draw a line through them, so that they don't distract you from the ones you're trying to concentrate on.

In questions with only one right answer, you can sometimes eliminate foils by recognizing two or more that are synonymous. Consider this question:

____ are blood cells that transform into macrophages. (a) Erythrocytes; (b) Platelets; (c) Neutrophils; (d) Red blood cells; (e) Monocytes

If you know that *erythrocytes* and *red blood cells* are the same thing, then obviously both *a* and *d* must be wrong, because if one were correct, they would both be correct, and this is against the rules of the test. Thus, you can immediately narrow the options down to *b, c,* and *e*. If you know that platelets are blood clotting elements (and, in fact, they are not blood cells at all), then you can further narrow the answer down to *c* or *e*, and if you've studied well, you'll know the correct answer is *e*.

Answering Matching Questions: Matching questions are just a variation on the multiple choice format. Be aware of the test rules, especially whether an item in the answer list can be used more than once. If not, and if you know one of them is the right answer to one question, then cross it out so that it doesn't distract you from picking the right answer to the next one.

Answering True/False Questions: Statements can be worded in subtle ways that make them false, so you must read such questions very carefully. Be especially careful of absolute expressions, such as *always* and *never*. For example, if a question says "Hypertension is always due to an excessive volume of blood," the word *always* warns you to think very carefully about this. True, an excessive blood volume is the most common cause of hypertension, but if you can think of even one exception, then the statement is false. If you can explain why a statement is false, or correct it, you will be doing more than mere guessing. The true/false section in this study guide gives you more practice in this technique.

Be careful of confusing cause and effect. If you are under pressure and prone to answer without thinking carefully, you might be inclined to answer *true* to a statement such as "Aneurysms are a common cause of hypertension," because you have come to associate the concepts of hypertension and aneurysm in your mind. But which is the cause and which is the effect? In reality, aneurysms are a result of hypertension, not the cause. The statement is false.

As a rule of thumb, assume that there are probably roughly equal numbers of true and false statements on a test (but be prepared for the idiosyncrasies of individual instructors). If a statement seems true and you've exercised reasonable care to look for anything that might make it false, then don't be afraid to answer *true*. Not every question is a trap.

Answering Essay Questions: The first rule in answering essay questions is to **read the question carefully.** This is true of all types of questions, but especially so for the true/false and essay formats. The consequences of not reading carefully can be extremely disappointing.

Pay close attention to such words as *describe* versus *explain.* If a question asks you to describe the change in blood pH that occurs during respiratory arrest, for example, you might scarcely need to do more than say a person stops breathing and the pH of the blood goes down. But if you are asked to explain it, then you need to discuss the mechanism that links the respiratory arrest to the pH change—carbon dioxide accumulates in the blood, it reacts with water to produce carbonic acid, carbonic acid releases hydrogen ions, and hydrogen ions lower the pH. If a question asks you to *compare and contrast* two things, you must describe their similarities *and* their differences. To do only one or the other risks getting only half the credit for the question.

Don't simply start writing without a clear idea of where you are headed. Someplace other than in the answer space (perhaps on the back of the test), write out a quick, sketchy list of the points you're going to make and the order in which you're going to present them. Think about how you're going to begin and close your essay, as well as the major things you need to say in between.

Try to express the general gist of your answer, or at least the most important part, in your first sentence; then elaborate on it or explain your point. For example, if the question is "Describe the physiological changes in the uterus from the time of ovulation to the time of menstruation," don't simply start out, "The corpus luteum secretes progesterone, which causes the uterus to . . ." Rather, you might write "The changes that occur in the uterus following ovulation essentially prepare it for a possible pregnancy and are controlled mainly by the corpus luteum and progesterone." Then you can state in more detail what the corpus luteum is, say what progesterone does, and wind up with what happens (menstruation) when pregnancy does not occur. When you come to a major transition in an essay, start a new paragraph—for example, "If pregnancy does not occur, however, . . ."

Diagrams often enhance an essay, and in many cases your answer should include them. If you were asked to describe the stages of oogenesis (egg formation in the ovaries), for example, a diagram of the cell divisions, labeled with the names of each stage, would get the point across more clearly and quickly than a couple of paragraphs of text. Many physiological questions are best answered by flowcharts—for example, if you were asked to explain a negative feedback loop that corrects for low blood pressure. Before the exam, ask your instructor if you have questions about how to answer his/her essays and if he/she accepts a diagram or flowchart in place of a written answer.

If the question is mathematical (for example, calculation of tests of kidney function), it would be a good idea to work out your answer on the back of the test or somewhere else, then copy it neatly to the answer space. In copying it, clearly show the logical order of the steps you followed. Clearly identify your answer, typically at the bottom right, perhaps emphasizing it by underlining or boxing it. Finally, don't omit the units of measurement. If you are calculating a glomerular filtration rate for the kidney, a number, such as 125, all by itself doesn't mean much. What is that—gallons per day? An answer such as 125 mL/min, by contrast, means something.

If a series of points are to be made, present them in the form of a numbered list. This makes it very easy to grade and enables the instructor to see the logic of your thinking. You are likely to get a better grade if your writing is easy to read and your reasoning is easy to follow. For example, if you were asked to explain how the kidneys can raise blood pressure by secreting renin, you could begin

1. The kidneys secrete renin.
2. Renin converts the plasma protein angiotensinogen to angiotensin I.
3. Angiotensin-converting enzyme in the lungs converts this to angiotensin II.

Be careful not to go beyond what a question asks; keep your answer succinct and to the point. For example, if the question asks you to write about events of the menstrual cycle from the time of ovulation to the time of menstruation, don't write about events of the first two weeks leading up to ovulation. Confine your answer to what was asked. You may lose points for misstatements about things you weren't required to write in the first place, or an instructor might take off points simply for not following instructions. Know when to stop. Don't stick your neck out by writing about more than the question asked. It won't impress the instructor, and it may even annoy him or her to have to read through a lot of irrelevant content. It also cuts short time that you may need to spend on other parts of the exam.

Proofread the *question* when you have finished your answer. Does the question have multiple parts? Have you answered them all? Rereading your answer will also help you avoid repeated or omitted words, especially at the end of a line or at places where you may have paused in your writing. Check your grammar, spelling, and punctuation when you proofread. Poor sentence composition is symptomatic of a poor education or carelessness. Remember, recommendation forms for nursing schools and other allied health programs routinely ask the writer to assess the student's facility for written expression. Often, a students essay answers are the only thing instructors have to go on when evaluating this. It is well worth checking your writing not only for scientific accuracy but also for writing style.

Summing Up: By now you might be feeling a bit overwhelmed, especially if you have met the class a time or two. The main thing is NOT to be defeated before you even begin. You are now armed with a fair knowledge of HOW to tackle A&P. The suggestions made here are based on research and years of personal experience; try them. Think of this introduction as the "Instruction Manual for Success in A&P" and you will survive. One more word of advice: **Ask for help if you need it.** You have tremendous resources available to you, starting with your instructor, teaching assistants, tutors, and counseling and/or learning center, not to mention all the online help the author and publishers provide. Take advantage of these gifts.

How to Sabotage Yourself

If all of this sounds like too much work, and you're still determined to fail, just follow these few simple instructions.

1. Skip class, or if you do attend, arrive fashionably late.

2. Don't buy the book, or if you buy it, don't read it.

3. Don't bother studying if you have to be somewhere else in 20 minutes; that's not enough time to get anything done.

4. Big test coming up? Beat the stress by relaxing with friends, going out for a few beers, or hanging out in an Internet chat room. Be sure to complain to your chat room friends about how there's no way you can pass the test tomorrow.

5. Don't ask questions in class; you're probably the only one who doesn't know the answer, and everyone else will think you're stupid.

6. Don't visit the instructor in his or her office; instructors don't want to be bothered.

7. If you miss a class, trust your friends' notes to be complete and accurate.

8. Be sure to pull an all-nighter before the exam; you don't have time to sleep.

9. Don't strain your brain trying to answer questions in the book and *Student Study Guide* from memory. Look up the answers and fill them in. You can fool your friends into thinking you're really smart (as long as they don't see your test grade).

10. When you study with friends, have a good time—chat about things unrelated to A&P.

11. The time to begin studying for an exam is the day before the test. Four hours ought to be plenty.

12. When studying, read each chapter several times and read your lecture notes several times. The best way to tell if you know the material is if it starts to look very familiar.

13. When reading the book, highlight most of it. If it weren't important, it wouldn't be in the book.

14. Don't take notes in lecture. You can't get it all down, anyway, and it would be better just to sit back and listen.

15. When taking an exam, read each question just enough to get the general gist of it; then hurry up and write your answer. You won't have time to read all the questions in their entirety.

Part II

Practice Exams

Chapter

Introduction

This section contains a practice exam for each chapter as well as for Atlas A. Each exam has a number of questions in several formats: 15 sentence completion, 10 matching, 10 true/false, 25 multiple choice, 15 word origins, 5 "which one does not belong" questions, and, in most chapters, additional exercises of various types. Use this as a final check on how well you know a chapter after answering the questions in the textbook. Take these exams with no notes or books in front of you, relying only on what you have learned.

Answer keys for these exams begin on page 211. Don't cheat yourself by yielding to the temptation to look ahead to the key before you have completed the test. No one is watching over your shoulder, criticizing your mistakes, or grading you on this. Do the best you can, learn from your errors, and repeat the exam if you do not do well the first time.

A. Short Answer: Complete each sentence or otherwise answer each question in the spaces provided or on another sheet of paper.

B. Matching: Match each of the 10 statements to one of the 24 answers. Each question has only one right answer, and answers are used only once.

C. True or False: Choose the statements that are incorrect and explain WHY they are wrong. This will help you far more than merely guessing. Your ability to answer WHY will tell you how thoroughly prepared you are for an exam.

D. Multiple Choice: Questions 1–20 are standard five-answer multiple choice and should not present a problem. Questions 21–25 are "multiple true/false." Note that the answer combinations are always the same throughout this study guide. In some cases, the statements are related to one topic, but in others, they are not. These questions are perhaps more challenging than regular multiple choice. However, if you treat each statement as a true/false question and look for one of the patterns in the answers (a–e), you can master them. If your answers do not fit one of the patterns, you need to reevaluate your choices. Be certain to read the introduction to this study guide and pay particular attention to the section on answering true/false questions.

E. Word Origins: This exercise is meant to develop your insight into biomedical vocabulary through recall of familiar word roots, prefixes, and suffixes. The first 10 ask you true/false questions about words found in the chapter. For the last 5, write the meaning of the root, prefix, or suffix and an example of a word from the chapter that uses that element.

F. Which One Does Not Belong: Read ALL the choices BEFORE choosing one of them. Students have a tendency to think the first item (a) is somehow the key to answering these questions; it is not. Some of the questions are very obvious; others are not so easy. A few may seem to have more than one right answer, but keep trying. The answer key has explanations to help you, but DON'T peek. At this stage in your preparation for an instructor's exam, you need to be challenged!

G. Figure Exercise/Matching: This section varies somewhat from chapter to chapter. Specific directions are given with each exercise. You will find that most questions over the diagrams are not generally labeling exercises. Instead, they challenge you beyond something you might have done in ninth or tenth grade. Presumably, you should be familiar with anatomy from studying your lab material, so the questions will ask you to put some concepts together with the structures.

1   Major Themes of Anatomy and Physiology

A. Short Answer

1. ___ is the study of structures of the body that are visible to the naked eye.

2. ___ is the tendency of the body to maintain an internal state of dynamic equilibrium.

3. The "if-then" prediction one makes from a hypothesis is called a(n) ___.

4. A(n) ___ is a group of individuals as similar as possible to an experimental group but who do not receive the experimental treatment.

5. ___ is the process by which a scientist must obtain approval of other experts in the field to get research funding and to get the findings published.

6. A statement or set of statements based on a large body of facts, laws, and confirmed hypotheses is called a scientific ___.

7. ___ wrote the first documented book proposing evolution by means of natural selection.

8. The structure and function of an organism result from the ___ to which the population adapted in the course of its evolution.

9. Homeostasis is normally maintained by cycles of self-correcting physiological responses called ___.

10. An organ is an anatomical structure made of at least two kinds of ___.

11. Structures within a cell that carry out specific functions for it are called ___.

12. A structure that carries out the body's ultimate response to a stimulus is called a(n) ___.

13. ___ is the separation of wastes from the tissues and their elimination from the body.

14. A(n) ___ suggests a method for answering a question.

15. Primates are defined in part by the presence of a(n) ___ thumb, which makes the hand as a whole ___, or able to grasp things by encircling them.

B. Matching

A. Aristotle
B. prognosis
C. Hominidae
D. Avicenna
E. hypothesis
F. dissection

G. selection pressures
H. Maimonides
I. holism
J. theory
K. *Homo erectus*
L. data

M. variables
N. placebos
O. Galen
P. ultrastructure
Q. vitalism
R. law of nature

S. *Australopithecus*
T. comparative anatomy
U. adaptation
V. diagnosis
W. Mammalia
X. reductionism

1. Oldest known bipedal primates

2. The family to which all living and extinct bipedal primates belong

3. Disease, predators, and competition, for example

4. Combined the findings of Aristotle and Galen with original discoveries; wrote *The Canon of Medicine*

5. *Of the Parts of Animals*

6. Considered only a speculation until it is tested by experiment or observation

7. Any factors that may affect the outcome of an experiment

8. Aids to distinguishing between pharmacological and psychosomatic effects

9. Theory in which organisms can be understood by the study of their component systems

10. The predicted outcome of an illness

C. True or False

1. The dissection of cadavers in medical school was forbidden in Vesalius' time.

2. Leeuwenhoek's simple microscopes possessed greater magnification than did Hooke's compound microscopes.

3. The level of organization between organ and cell is the organelle.

4. A baroreflex is an example of a control center.

5. Negative feedback is usually a health-maintaining process.

6. Positive feedback always represents a threat to one's health or life, and it must be prevented or stopped if possible.

7. Color vision is rare among mammals other than primates.

8. The first person to observe blood capillaries was William Harvey.

9. A statement such as "The human forearm has two bones, with the radius located lateral to the ulna" is a product of the inductive method of science.

10. The steadily deteriorating condition of the brain resulting from a high fever is an example of positive feedback.

D. Multiple Choice

1. The study of the structure of individual cells is called: (a) histology; (b) cytology; (c) auscultation; (d) percussion; (e) microscopic anatomy.

2. All living organisms possess all of the following properties *except:* (a) metabolism; (b) breathing; (c) excitability; (d) reproduction; (e) excretion.

3. Homeostasis is maintained primarily by: (a) negative feedback; (b) positive feedback; (c) feedback inhibition; (d) allosteric inhibition; (e) autoregulation.

4. Palpation is the study of anatomy by means of: (a) dissection; (b) the electron microscope; (c) the inductive method; (d) touch; (e) sound.

5. Pathophysiology is the study of: (a) how different species compare; (b) human development; (c) disease processes; (d) the nervous system; (e) the digestive system.

6. An expert on ultrastructure probably would know the most about: (a) gross anatomy; (b) tissue structure; (c) organelles; (d) organ systems; (e) biochemistry.

7. The development of unspecialized cells into more specialized ones is: (a) differentiation; (b) assimilation; (c) anabolism; (d) reproduction; (e) growth.

8. All animals are heterotrophic, meaning: (a) they have true nuclei; (b) they have a variety of tooth forms; (c) they are endothermic; (d) they exhibit homeostasis; (e) they consume other organisms.

9. Which of the following is an example of selection pressure? (a) adaptation; (b) mutations; (c) bipedal locomotion; (d) climate; (e) endothermy

10. An informed conjecture that hasn't been tested yet is a: (a) theory; (b) hypothesis; (c) law of nature; (d) tentative fact; (e) principle.

11. The word *physiology* comes from ___ attempt to distinguish between natural and supernatural causes of things. (a) Galen's; (b) Vesalius'; (c) Aristotle's; (d) Hooke's; (e) Harvey's

12. The first person to produce a comprehensive, accurate atlas for the teaching of human anatomy was: (a) Hippocrates; (b) Galen; (c) Vesalius; (d) Van Leeuwenhoek; (e) Maimonides.

13. There usually is more verified scientific information in a(n) ___ than in any of these other concepts. (a) theory; (b) law; (c) fact; (d) hypothesis; (e) experiment

14. The arboreal habits of the ancestors of humans can account for all of the following characteristics of *Homo sapiens except:* (a) stereoscopic vision; (b) a prehensile thumb; (c) physiological variation; (d) color vision; (e) a highly flexible shoulder.

15. Precision in the language of science is important because small changes in terms may cause large changes in meanings. For example, which of these would you expect your surgeon to know well if you needed an intestinal operation? (a) ileum; (b) illeum; (c) ilium; (d) illeum; (e) iliac

16. Auscultation is the study of the human body by means of: (a) touching; (b) listening; (c) dissection; (d) noninvasive medical imaging; (e) controlled experimentation.

17. Which of these is used to best visualize brain function? (a) angiography; (b) sonography; (c) fMRI; (d) CT; (e) PET

18. Humans exhibit heterodonty, meaning: (a) they walk upright; (b) their cells have true nuclei; (c) they must eat other organisms to live; (d) they have a hollow central nervous system; (e) they have varied teeth.

19. Anatomical terms coined from the names of people are: (a) pseudonyms; (b) nominae; (c) synonyms; (d) acronyms; (e) eponyms.

20. Even though physicians and medical professors treated his book as unquestionable dogma for 1,500 years, ___ warned people to trust their own observations more than what they read in it. (a) Hippocrates; (b) Galen; (c) Avicenna; (d) Paracelsus; (e) Vesalius

21. Which of these is/are true about evolution?
    1. Individuals can evolve throughout a lifetime.
    2. A requirement is mutation in DNA structure.
    3. Selection pressures, such as disease and predators, weaken an individual's chance for evolution.
    4. The presence of vestigial structures in humans provides some evidence for human evolution.

    (a) 1 & 3; (b) 2 & 4; (c) 1, 2, & 3; (d) 4 only; (e) all the above

22. Which of these is/are true?
    1. There are cases in humans where some organs are reversed from the normal position in the body.
    2. Physiological parameters, such as blood pressure, tend to remain constant throughout one's lifetime.
    3. Anabolic processes occur during growth of an individual.
    4. A "reference" man is based on findings in "average" 40-year-old men.

    (a) 1 & 3; (b) 2 & 4; (c) 1, 2, & 3; (d) 4 only; (e) all the above

23. Which of these is/are true about homeostasis?
    1. It can be described as dynamic equilibrium.
    2. It is maintained by negative feedback loops.
    3. It is a tendency to maintain a stable internal environment.
    4. It requires a sensor, an effector, and a regulatory mechanism.

    (a) 1 & 3; (b) 2 & 4; (c) 1, 2, & 3; (d) 4 only; (e) all the above

24. Which of these is/are true?
    1. Human anatomy textbooks illustrate and describe what is "normal" for about 70% of people.
    2. In order to maintain constant body temperature, we can either dilate or constrict blood vessels.
    3. Blood pressure is partially controlled by self-correcting negative feedback loops.
    4. Positive feedback normally results in a disease state.

    (a) 1 & 3; (b) 2 & 4; (c) 1, 2, & 3; (d) 4 only; (e) all the above

25. Which of these is/are true about the language of science?
    1. It can be understood better by taking words apart into their component parts.
    2. A word root is used to modify the core meaning of the word.
    3. The plural of *foramen* is *foramina.*
    4. The plural of *brachium* is *bronchi.*

    (a) 1 & 3; (b) 2 & 4; (c) 1, 2, & 3; (d) 4 only; (e) all the above

E. Word Origins

1. In *anatomy, -tomy* means "study."
2. In *heterotrophic, hetero-* means "other."
3. In *eukaryotic, eu-* means "true."
4. In *endothermic, endo-* means "warm."
5. In *Mammalia, mamma-* means "breast."
6. In *cytology, cyto-* means "cell."
7. In *histology, histo-* means "organ."
8. In *arboreal, arbor-* means "climbing."
9. In *bipedal, ped-* means "foot."
10. In *homeostasis, stasis-* means "to stay."
11. dis-
12. homeo-
13. physio-
14. -sect
15. cadere-

F. Which One Does Not Belong?

1. (a) camouflage; (b) disease; (c) competition; (d) climate

2. (a) color vision; (b) prehensile thumb; (c) stereoscopic vision; (d) bipedalism

3. (a) reductionism; (b) vestigial organ; (c) natural selection; (d) sliding filament

4. (a) Hippocrates; (b) Walter Cannon; (c) Robert Hooke; (d) Claude Bernard

5. (a) carcinomata; (b) viscus; (c) calyx; (d) axilla

Atlas A   General Orientation to Human Anatomy

A. Short Answer

1. When the forearm is ___, the radius and ulna cross each other and the palm is turned downward or to the rear.

2. The ___ plane is one that divides the body as evenly as possible into right and left halves.

3. If structure A is closer to the body surface than structure B, we say A is ___ to B and B is ___ to A.

4. The rostral end of an animal is toward the ___, while the caudal end is toward the ___.

5. The armpit is anatomically known as the ___ region and the pit on the front of the elbow is called the ___ region.

6. The wrist is called the ___ region and the ankle is the ___ region.

7. If the abdomen is divided into a 3x3 grid, the square surrounding the navel is called the ___ region and the square immediately inferior to that is the ___ region.

8. On the same 3x3 grid, the upper row consists of the ___ region in the middle and the ___ on each side of that.

9. The thoracic cavity is subdivided by a medial wall, the ___, with a ___ cavity to the right and left of it.

10. The pleurae and pericardium consist of two layers, the outer ___ layer and inner ___ layer, with a fluid-filled space between them.

11. The ___ of the hands and feet are identified by the presence of nails.

12. The two organ systems concerned with internal communication and coordination are the ___ and ___ systems.

13. The two organ systems concerned with fluid transport or circulation are the ___ and ___ systems.

14. The three organ systems concerned with the intake and output of substances are the digestive, ___, and ___ systems.

15. If structure A is farther away from a point of origin or attachment than structure B is, we say A is ___ to B and B is ___ to A.

B. Matching

A. supine	G. prone	M. lateral	S. median
B. coronal	H. transverse	N. anterior	T. dorsal
C. distal	I. superior	O. brachium	U. antebrachium
D. carpus	J. thoracic	P. abdominal	V. parietal
E. visceral	K. lumbar	Q. peritoneum	W. cranial
F. vertebral	L. mesentery	R. coelom	X. meninges

1. Position of a human body when lying on its anterior surface

2. Plane that would separate the sternum from the spinal column

3. The same as ventral, in humans

4. Position of the patellar region relative to the coxal region

5. Region from elbow to wrist

6. Region of the lower back

7. Membranous linings of the cranial cavity

8. Serous membrane that suspends the intestines from the abdominal wall

9. Serous membrane that lines the abdominopelvic cavity

10. Body cavity that contains the brain

C. True or False

1. The sternum is inferior to the heart.

2. It is impossible for a median section of the body to show both eyes.

3. In anatomical position, the radius and ulna are parallel.

4. There can be only one true parasagittal plane of the body.

5. A cross section through the heart would show all four chambers.

6. Each organ belongs to one and only one organ system.

7. The heart and lungs are situated within the pleural cavity.

8. The greater omentum hangs from the inferolateral border of the stomach and covers the intestine.

9. The immune system is not an organ system, but it defends the body against invading pathogens.

10. Viscera superior to the diaphragm cannot be described as retroperitoneal.

D. Multiple Choice

1. The spinal column can be described as ___ to the ribs because it is closer to the midline of the body. (a) proximal; (b) medial; (c) superficial; (d) lateral; (e) parietal

2. The heart is ___ to the diaphragm. (a) superior; (b) posterior; (c) distal; (d) superficial; (e) proximal

3. Which of these does *not* belong with the rest? (a) endocrine; (b) epithelium; (c) muscular; (d) skeletal; (e) circulatory

4. The ___ is divided into the RUQ, RLQ, LUQ, and LLQ. (a) body cavity; (b) brain; (c) abdomen; (d) back; (e) body as a whole

5. The appendix normally lies in the: (a) pleural cavity; (b) right lower quadrant; (c) left lower quadrant; (d) pelvic cavity; (e) gastric region.

6. Which of these is *not* one of the body cavities? (a) pericardial cavity; (b) pelvic cavity; (c) epigastric cavity; (d) vertebral canal; (e) pleural cavity

7. The uterine cavity is: (a) retroperitoneal; (b) a potential space; (c) posterior to the umbilicus; (d) in the abdominal cavity; (e) posterior to the spine.

8. The diaphragm lies approximately on a ___ plane of the body. (a) frontal; (b) lateral; (c) midsagittal; (d) coronal; (e) transverse

9. The urinary bladder is ___ to the external genitalia. (a) superior; (b) inferior; (c) lateral; (d) superficial; (e) distal

10. The carpal region is ___ to the axillary region. (a) distal; (b) medial; (c) posterior; (d) proximal; (e) peripheral

11. The ___ region is inferior to the lateral abdominal region. (a) inguinal; (b) hypogastric; (c) umbilical; (d) hypochondriac; (e) epigastric.

12. The liver lies mostly in which region of the abdomen? (a) left upper quadrant; (b) right upper quadrant; (c) umbilical; (d) right lateral abdominal; (e) right inguinal

13. The ___ layer of the pleura lies against the inside of the rib cage. (a) posterior; (b) lateral; (c) visceral; (d) peripheral; (e) parietal

14. The ___ lies in the mediastinum. (a) brain; (b) spinal cord; (c) trachea; (d) liver; (e) urinary bladder

15. The external surfaces of the stomach and intestines are covered by a/an: (a) pleura; (b) serosa; (c) mesocolon; (d) omentum; (e) meninx.

16. When you hold out your hand for someone to place money in your palm, you are holding it in the ___ position. (a) supinated; (b) anatomical; (c) abducted; (d) pronated; (e) transverse

17. A cross section of an organ is a cut along its ___ plane. (a) transverse; (b) sagittal; (c) coronal; (d) lateral; (e) frontal

18. The brachial region is the: (a) thigh; (b) lower back; (c) chest; (d) neck; (e) arm.

19. The popliteal region is the: (a) pit of the elbow; (b) armpit; (c) back of the neck; (d) back of the knee; (e) ankle.

20. The ___ is a retroperitoneal viscus of the abdominopelvic cavity. (a) stomach; (b) liver; (c) kidney; (d) small intestine; (e) spleen

21. Which of these is/are located in the appendicular region?
    1. crural region
    2. leg
    3. cubital region
    4. trunk
    (a) 1 & 3; (b) 2 & 4; (c) 1, 2, & 3; (d) 4 only; (e) all the above

22. The true statement(s) about directional relationships is/are:
    1. The carpals are proximal to the axillary region.
    2. The umbilical region is inferior to the inguinal region.
    3. The lumbar region is lateral to the vertebral region.
    4. The integumentary system is superficial to the other systems.
    (a) 1 & 3; (b) 2 & 4; (c) 1, 2, & 3; (d) 4 only; (e) all the above

23. Which of these is/are correct match(es) between systems and their organs?
    1. endocrine system—thymus gland
    2. nervous system—thyroid gland
    3. lymphatic system—spleen
    4. respiratory system—blood vessels
    (a) 1 & 3; (b) 2 & 4; (c) 1, 2, & 3; (d) 4 only; (e) all the above

24. The mediastinum is occupied by the:
    1. lungs.
    2. esophagus.
    3. viscera.
    4. heart.
    (a) 1 & 3; (b) 2 & 4; (c) 1, 2, & 3; (d) 4 only; (e) all the above

25. The abdominal cavity contains the:
    1. urinary bladder.
    2. stomach.
    3. uterus.
    4. liver.
    (a) 1 & 3; (b) 2 & 4; (c) 1, 2, & 3; (d) 4 only; (e) all the above

E. Word Origins

1. In *parasagittal*, *para-* means "two."
2. In *subcostal*, *costa-* means "margin."
3. In *intertubercular*, *inter-* means "between."
4. In *hypochondriac*, *chondri-* means "illness."
5. In *antebrachial*, *ante-* means "against."
6. In *antebrachial*, *brachi-* means "arm."
7. In *pleural*, *pleura-* means "lung."
8. In *pericardium*, *cardi-* means "heart."
9. In *retroperitoneal*, *retro-* means "backwards."
10. In *mesentery*, *enter-* means "inside."
11. epi-
12. hypo-
13. peri-
14. sub-
15. gastro-

F. Which One Does Not Belong?

1. (a) proximal; (b) deep; (c) inferior; (d) coronal

2. (a) palmar; (b) plantar; (c) antebrachial; (d) carpal

3. (a) pleura; (b) pericardium; (c) meninges; (d) peritoneum

4. (a) immune system; (b) respiratory system; (c) urinary system; (d) digestive system

5. (a) kidneys; (b) stomach; (c) adrenal gland; (d) abdominal aorta

2   The Chemistry of Life

A. Short Answer

1. Large molecules made of many similar or identical subunits, such as starch and protein, are called ___.

2. The chemical properties of an element are determined solely by its ___, found in the outermost energy level of the atom.

3. A positively charged ion is called a(n) ___ and a negatively charged ion is called a(n) ___.

4. Antioxidants, such as vitamin E, protect the body from the harmful effects of uncharged particles called ___.

5. Molecules that reversibly bind to a protein, such as a hormone to its receptor, are called ___.

6. Bonds between atoms formed by the sharing of electrons are called ___ bonds.

7. Water molecules cling together and many large molecules hold their three-dimensional shapes due to weak attractions called ___.

8. Triglycerides consist of three ___ covalently bonded to glycerol.

9. A chemical reaction that consumes water to break a covalent bond in a larger molecule is called ___.

10. ___, such as Ca^{2+}, are inorganic components of enzymes that can induce folding of the active site, thus activating the enzyme.

11. A solution with a pH of 6 is how many times more acidic than a solution with a pH of 9?

12. A molecule becomes ___ when electrons are removed from it.

13. The place where a substrate binds to an enzyme is called the enzyme's ___.

14. A fatty acid with numerous C=C bonds is said to be ___.

15. Enzymes called ___ add phosphate groups to other organic molecules.

B. Matching

A. lipids	G. polar covalent	M. atomic weight	S. protein
B. dehydration synthesis	H. ionic bond	N. anabolism	T. atomic mass unit
C. colloid	I. cation	O. moiety	U. monosaccharides
D. double covalent	J. isotope	P. proteoglycans	V. electron acceptor
E. hydrogen ion	K. hydrolysis	Q. base	W. molecular weight
F. ATP	L. substrates	R. disaccharides	X. phospholipids

1. Universal energy transfer molecule

2. A free proton

3. One-twelfth the mass of a ^{12}C atom

4. Hydrophobic organic compounds

5. Sucrose and lactose

6. Amphiphilic molecules that form a major part of cell membranes

7. Built from amino acids

8. Reaction that combines small, low-energy molecules into a larger, higher-energy molecule

9. Produces water as a by-product

10. Bond in which electrons are unequally shared between two nuclei

C. True or False

1. The products of catabolism contain more free energy than the reactants.

2. One mole of a chemical always contains the same number of molecules, no matter what the chemical is.

3. Gamma rays are the most penetrating form of ionizing radiation.

4. Fats are described as saturated if they have no double bonds between their carbon atoms.

5. Cholesterol does the body more good than harm.

6. The biological half-life of a radioisotope is usually longer than its physical half-life.

7. Polysaccharides over 100 monomers long are classified as proteins.

8. Trans-fatty acids are easier to digest than cis-fatty acids.

9. To be considered a molecule, a particle must contain atoms of two or more different elements.

10. Oxidation reduces the free energy content of a molecule.

D. Multiple Choice

1. Water has a surface tension because of its: (a) hydrophilic character; (b) hydration spheres; (c) valence; (d) hydrogen bonds; (e) heat capacity.

2. Any chemical reaction that requires an energy input is said to be: (a) endergonic; (b) catabolic; (c) metabolic; (d) exergonic; (e) intrinsic.

3. The breakdown of large molecules into smaller ones is: (a) condensation; (b) catabolism; (c) anabolism; (d) metabolism; (e) endergonic.

4. The most abundant element in the earth's crust is oxygen. The most abundant one in the body, by weight, is: (a) carbon; (b) hydrogen; (c) nitrogen; (d) water; (e) also oxygen.

5. The average person in the United States is exposed to about ___ of background radiation per year. (a) 5 Sv; (b) 140 Sv; (c) 3.6 mSv; (d) 0.6 Sv; (e) 600 mSv

6. The number of protons in the nucleus of an element is indicated by its: (a) atomic number; (b) mass number; (c) isotope number; (d) atomic weight; (e) molarity.

7. The opposite of a condensation reaction is: (a) oxidation; (b) reduction; (c) synthesis; (d) hydrolysis; (e) anabolism.

8. Enzymes act by reducing the ___ of a chemical reaction. (a) entropy; (b) probability; (c) temperature; (d) energy yield; (e) activation energy

9. A base is any chemical that: (a) accepts H^+; (b) produces H^+; (c) produces OH^-; (d) accepts OH; (e) has a $[H^+] > 10^{-7}$ M.

10. The arrangement of a polypeptide into a fibrous or globular shape is called its: (a) primary structure; (b) secondary structure; (c) tertiary structure; (d) quaternary structure; (e) denatured structure.

11. A true statement about ions and free radicals is: (a) both are produced by normal metabolic reactions in the body; (b) a free radical has a charge of only ±1, whereas an ion can have charges anywhere from ±1 to ±3; (c) a free radical can oxidize other molecules but an ion cannot; (d) an ion has only one atomic nucleus, whereas a free radical can be a large molecule; (e) ions are positive and free radicals are negative.

12. Two different chemicals with the same number and types of atoms arranged in different ways are called: (a) isomers; (b) isobars; (c) isoclines; (d) isotopes; (e) isoenzymes.

13. Glucose, fructose, and galactose are: (a) polysaccharides; (b) disaccharides; (c) monosaccharides; (d) polypeptides; (e) oligopeptides.

14. All the body's steroids, such as estrogen and testosterone, are synthesized from: (a) triglycerides; (b) amino acids; (c) fatty acids and glycerol; (d) cholesterol; (e) prostaglandins.

15. Salts, acids and bases that ionize in body fluids are called: (a) free radicals; (b) isotopes; (c) ionizing radiation; (d) electrolytes; (e) isomers.

16. Pepsin, amylase, and trypsin differ in that: (a) they are an oligopeptide, polypeptide, and protein, respectively; (b) one is saturated, one is monounsaturated, and one is polyunsaturated; (c) they function best at three different temperatures; (d) they function best at three different pHs; (e) they are an enzyme, a coenzyme, and an enzyme substrate, respectively.

17. Which of these is *not* one of the four primary classes of biological macromolecules? (a) enzymes; (b) lipids; (c) nucleic acids; (d) proteins; (e) carbohydrates

18. The isotopes ^{12}C, ^{13}C, and ^{14}C all behave the same chemically because they: (a) have the same number of neutrons; (b) have the same number of electrons; (c) have equal energy; (d) have the same atomic mass; (e) are all radioactive.

19. A 1 millimolar solution of an ion with a valence of $^+2$ can also be said to have a concentration of: (a) 1 M; (b) 2 mEq/L; (c) 2 mM/L; (d) 1 Eq/L; (e) 2 millimoles/liter.

20. Other than hydrogen bonds, which of these is most important in stabilizing tertiary protein structure? (a) polar covalent bonds; (b) van der Waals forces; (c) electron energy; (d) ionic bonds; (e) covalent forces

21. Examples of protein functions include:
    1. They can act as membrane channels and carriers.
    2. They can be enzymes that catalyze reactions.
    3. They can bind cells together.
    4. They can act as chemical messengers.

    (a) 1 & 3; (b) 2 & 4; (c) 1, 2, & 3; (d) 4 only; (e) all the above

22. Which of these is/are true about enzymes?
    1. Isoenzymes can be used to diagnose certain diseases.
    2. Most are consumed by the reaction they catalyze.
    3. They exhibit substrate specificity.
    4. They require activation energy to function properly.

    (a) 1 & 3; (b) 2 & 4; (c) 1, 2, & 3; (d) 4 only; (e) all the above

23. Which of these is/are true about ATP?
    1. It is short-lived and therefore has limited usefulness in living things.
    2. Enzymatically splitting ATP yields heat, energy to do work, ADP, and P_i.
    3. More ATP is produced in fermentation than in aerobic respiration.
    4. Oxidation of glucose produces ATP.

    (a) 1 & 3; (b) 2 & 4; (c) 1, 2, & 3; (d) 4 only; (e) all the above

24. Which of these is/are true?
    1. Free radicals are normally not harmful to living tissue.
    2. Anabolic-androgenic steroids like Dianabol made after 1991 are much safer than earlier forms.
    3. Irene Joliot-Curie was the first woman to earn a Nobel Prize.
    4. Ion trapping or pH partitioning can be beneficial in ridding the body of some poisons.

    (a) 1 & 3; (b) 2 & 4; (c) 1, 2, & 3; (d) 4 only; (e) all the above

25. Glucose, cellulose, maltose, and glycogen are all:
    1. sugars.
    2. starches.
    3. polysaccharides.
    4. carbohydrates.

    (a) 1 & 3; (b) 2 & 4; (c) 1, 2, & 3; (d) 4 only; (e) all the above

E. Word Origins

1. In *atom*, *tom* means "small."
2. In *isotope*, *iso-* means "same."
3. In *calorie*, *calor-* means "fat."
4. In *colloid*, *-oid* means "resembling."
5. In *exergonic*, *erg-* means "heat."
6. In *glycogen*, *-gen* means "starch."
7. In *anabolism*, *ana-* means "up."
8. In *monosaccharide*, *sacchar-* means "sugar."
9. In *isomer*, *-mer* means "the sea."
10. In *hydrolysis*, *-lysis* means "splitting apart."
11. poly-
12. -jug
13. glyco-
14. amphi-
15. cata-

F. Which One Does Not Belong?

1. (a) glycogen → glucose; (b) peptide → amino acids; (c) triglyceride → fatty acids; (d) steroids → proteoglycans

2. (a) glycerol; (b) prostaglandin; (c) glycogen; (d) eicosanoid   Lipids

3. (a) gynomastia; (b) sterility; (c) obesity; (d) coronary artery disease

4. (a) reversible; (b) synthesis; (c) exchange; (d) decomposition

5. (a) cellulose; (b) glycogen; (c) glucose; (d) starch

3   Cellular Form and Function

A. Short Answer

1. The most important quality of a good microscope is not magnification but ___, the ability to distinguish details.

2. A cell with numerous pointed processes is described as ___ in shape because of its resemblance to a star.

3. All contents of a cell between the plasma membrane and the nuclear envelope are called the ___.

4. Phospholipid molecules with carbohydrate groups covalently bound to them are called ___.

5. Proteins that extend all the way through a plasma membrane are called ___ proteins.

6. Channel proteins that allow water to pass into/out of cells are called ___.

7. On some cells, numerous closely spaced microvilli form a fringe called the ___ that serves to increase the cell's surface area.

8. Macromolecules can move into and out of the cell nucleus through passages called ___.

9. The immune system distinguishes the body's own cells from foreign cells by the composition of a fuzzy carbohydrate cell coat called the ___.

10. ___ produces a sodium gradient across a cell membrane that acts as a source of potential energy for the cell.

11. Osmosis is driven by the concentration of ___ sequestered on one side of a selectively permeable membrane.

12. In the process called ___, molecules in the extracellular fluid bind to receptors on the cell surface and are taken into the cell in a membrane-bounded vesicle.

13. The space between the membranes of rough or smooth endoplasmic reticulum is called a ___.

14. Liver and kidney cells have numerous ___, that neutralize reactive molecules that could otherwise cause cell damage.

15. The ___ is an ATP-consuming antiport system that pumps Na^+ out of a cell and K^+ into it.

B. Matching

A. fusiform	G. pinocytosis	M. cytoskeleton	S. active transport
B. microfilaments	H. microvilli	N. squamous	T. transport maximum
C. cilia	I. peroxisomes	O. axoneme	U. smooth ER
D. SEM	J. hypertonic	P. glycoproteins	V. channel protein
E. hypotonic	K. cuboidal	Q. saturated	W. facilitated diffusion
F. intermediate filaments	L. TEM	R. G proteins	X. plasma membrane

1. Collection of proteins that determine cell shape

2. Flat and scale-like in shape

3. Produces high-resolution, three-dimensional images

4. Described by the fluid-mosaic model

5. Membrane proteins involved in activating second messenger systems

6. Movement of a solute up its concentration gradient with the expenditure of ATP

7. Occurs when membrane carriers are saturated with solute

8. The microtubular core of a cilium

9. Provide strength to hair and fingernails

10. Describes a solution in which cells would tend to absorb water

C. True or False

1. The nuclear lamina, a web of protein filaments under the nuclear envelope, provides attachments for chromatin and helps regulate the cell cycle.

2. Cells use exocytosis mainly to get rid of their waste products.

3. The cell walls of humans are composed of a fluid mosaic of proteins and phospholipids.

4. Fluid-phase pinocytosis takes in droplets of ECF without modifying its composition or concentration.

5. Facilitated diffusion does not consume ATP.

6. Microvilli can be recognized from their 9+2 internal arrangement of microtubules.

7. Carrier-mediated transport can occur in the absence of ATP.

8. Louis Pasteur believed in the spontaneous generation of living cells.

9. Mitochondrial DNA mutates at a much higher rate than does nuclear DNA.

10. Na^+–K^+ pumps are used to make energy— for example, when it is cold and the body needs more heat.

D. Multiple Choice

1. Most of the molecules in a plasma membrane are: (a) carbohydrates; (b) polypeptides; (c) glycoproteins; (d) phospholipids; (e) cholesterol.

2. Which of the following processes requires a membrane, but not necessarily a living one? (a) simple diffusion; (b) active transport; (c) osmosis; (d) pinocytosis; (e) exocytosis

3. All are functions of the glycocalyx *except:* (a) cell division; (b) cancer defense; (c) embryonic development; (d) protection; (e) immunity.

4. All of the following organelles have at least one unit membrane around them *except:* (a) centrioles; (b) mitochondria; (c) lysosomes; (d) Golgi vesicles; (e) smooth endoplasmic reticulum.

5. Resting membrane potential is continuously maintained by the action of: (a) endocytosis; (b) the Na^+–K^+ ATPase pump; (c) osmosis; (d) secondary active transport; (e) transcytosis.

6.   A type of lipid found in plasma membranes is: (a) peripheral; (b) glycocalyx; (c) glycoprotein; (d) pores; (e) cholesterol.

7. Clathrin-coated vesicles are formed in: (a) receptor-mediated endocytosis; (b) phagocytosis; (c) exocytosis; (d) the Na^+–K^+ pump; (e) the glycocalyx.

8. Red blood cells become crenated in a(n) ___ solution. (a) physiological saline; (b) hypertonic; (c) isotonic; (d) tonic; (e) hypotonic

9. Autolysis and autophagy are two functions of: (a) the nucleus; (b) lysosomes; (c) rough endoplasmic reticulum; (d) the Golgi complex; (e) mitochondria.

10. Leukocytes engulf bacteria by the process of: (a) phagocytosis; (b) pinocytosis; (c) exocytosis; (d) macrocytosis; (e) active transport.

11. If a poison, such as cyanide, suddenly halted a cell's ATP synthesis, ___ would cease. (a) osmosis; (b) metabolism; (c) simple diffusion; (d) facilitated diffusion; (e) active transport

12. Which of these is *not* a proposition of the modern cell theory? (a) All cells arise from other cells. (b) All cells are enclosed within a membrane. (c) The cell is the smallest unit of life. (d) The cells of all species are basically alike in chemical composition. (e) All functions of an organism are due to the functioning of its cells.

13. ___ is the means by which oxygen enters red blood cells. (a) Diffusion; (b) Osmosis; (c) Facilitated diffusion; (d) Pinocytosis; (e) Active transport

14. Factors that affect the rate of solute diffusion include all *except*: (a) membrane shape; (b) concentration gradient; (c) temperature; (d) availability of ATP; (e) molecular weight.

15. All are functions of cilia *except:* (a) to bind odor molecules; (b) to move mucus in the respiratory system; (c) balance; (d) to detect fluid flow in the kidneys; (e) to cushion and protect the plasma membrane.

16. Ribosomes can be found either on the rough endoplasmic reticulum or: (a) in the nucleus; (b) attached to the plasma membrane; (c) free in the cytosol; (d) on the smooth endoplasmic reticulum; (e) within microtubules.

17. A cuboidal cell with a diameter of 30µm on a side has a volume of ___ µm^3. (a) 9,000; (b) 900; (c) 270; (d) 27,000; (e) 3,000

18. Most knowledge of organelles resulted from the invention of the: (a) transmission electron microscope; (b) scanning electron microscope; (c) compound microscope; (d) CT scan; (e) PET scan.

19. A protein in the plasma membrane that opens when a chemical binds to it is called a: (a) second messenger; (b) ligand-gated channel; (c) receptor; (d) peripheral protein; (e) motor molecule.

20. Vesicles destined to release their contents from a cell by exocytosis are packaged by the: (a) lysosomes; (b) rough endoplasmic reticulum; (c) Golgi complex; (d) peroxisomes; (e) cytoskeleton.

21. The role(s) of membrane proteins is/are as:
    1. carriers.
    2. movement.
    3. receptors.
    4. second messenger systems.

    (a) 1 & 3; (b) 2 & 4; (c) 1, 2, & 3; (d) 4 only; (e) all the above

22. Adenylate cyclase:
    1. phosphorylates other enzymes.
    2. is an enzyme that acts on ATP to form cAMP.
    3. is a receptor for various ligands.
    4. catalyzes the formation of a common second messenger.

    (a) 1 & 3; (b) 2 & 4; (c) 1, 2, & 3; (d) 4 only; (e) all the above

23. Which of these is/are true about the Na$^+$–K$^+$ ATPase pump?
    1. It maintains greater Na+ concentration in the ECF.
    2. Half the calories we eat in a day go to fueling it.
    3. It aids glucose movement in secondary active transport.
    4. It produces body heat.

    (a) 1 & 3; (b) 2 & 4; (c) 1, 2, & 3; (d) 4 only; (e) all the above

24. Which of these is/are true?
    1. Calcium channel blockers lower blood pressure by preventing blood vessel smooth muscle from contracting.
    2. In cystic fibrosis no chloride pumps are made by cells of the respiratory system.
    3. Most cells in the human body have at least one cilium.
    4. Mitochondria are inherited equally from both maternal and paternal sources.

    (a) 1 & 3; (b) 2 & 4; (c) 1, 2, & 3; (d) 4 only; (e) all the above

25. Which of these is/are true?
    1. Glycolipids and glycoproteins form the glycocalyx.
    2. The contents of the ECF and ICF are normally the same.
    3. The rate of solute transport increases as solute concentration increases until T_m is reached.
    4. People with familial hypercholesterolemia can live normal life spans if they adhere to a strict low-fat diet.

    (a) 1 & 3; (b) 2 & 4; (c) 1, 2, & 3; (d) 4 only; (e) all the above

E. Word Origins

1. In *squamous, squam-* means "flat, scaly."
2. In *phagocytosis, phago-* means "drink."
3. In *fusiform, fusi-* means "joined, fused."
4. In *pseudopod, pseudo-* means "false."
5. In *microvillus, micro-* means "small."
6. In *glycocalyx, -calyx* means "cup."
7. In *cisterna, cistern-* means "space."
8. In *lysosome, -lyso* means "body."
9. In *osmosis, osmo-* means "water."
10. In *facilitated, facili-* means "easy."
11. dyn-
12. -plasm
13. proto-
14. -ous
15. -ton

F. Which One Does Not Belong?

1. (a) mitochondria; (b) microtubules; (c) endoplasmic reticulum; (d) Golgi complex

2. (a) filtration; (b) osmosis; (c) facilitated diffusion; (d) diffusion

3. (a) flagella; (b) cilia; (c) microtubules; (d) microvilli

4. (a) Na^+–K^+ ATPase pump; (b) maintenance of membrane potential; (c) regulation of cell volume; (d) cotransport

5. (a) hypercholesterolemia; (b) Kearns-Sayre syndrome; (c) mitochondrial myopathy; (d) Leber hereditary optic neuropathy

G. Figure Exercise

Match the structure with its description or function.

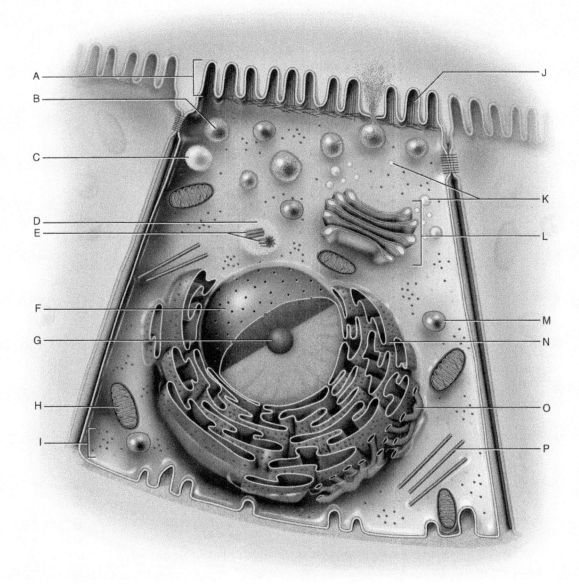

1. Made by Golgi complex; used by white blood cells to digest bacteria
2. Surrounded by double unit membrane; contains nucleoli and chromatin
3. Continuous with the outer membrane of the nuclear envelope; functions in making new cell membrane
4. Extensive in liver and kidney cells; detoxifies, synthesizes lipids; stores calcium
5. Assembles building blocks of amino acids into proteins according to the genetic code
6. Processes and packages proteins for exocytosis
7. Extensions of plasma membrane that increase surface area; have sensory function in some organs
8. Were once free, living bacteria; sites for ATP synthesis
9. Made of specific arrangement of microtubules; assemble the spindle for cell division
10. Hollow protein cylinders that make up the mitotic spindle

4   Genetics and Cellular Function

A. Short Answer

1. The "sense" portions of a pre-mRNA strand made in transcription are called ___.

2. Having two different alleles for the same trait is a(n) ___ condition.

3. About ___ of nuclear DNA houses the approximately 25,000 genes in the human genome; the rest is noncoding but is important in ___ and chromosome structure.

4. Alkaptonuria is an example of ___, in which one gene can have many phenotypic effects.

5. Skin color and eye color are examples of ____.

6. A nucleotide triplet on mRNA must bind to a complementary triplet on tRNA, called a(n) ___, for translation to occur.

7. ___ inhibit the development of cancer by various methods.

8. An allele is said to be ___ if it is not phenotypically expressed in the presence of another allele of the same gene.

9. Oncogenes may trigger cancer by stimulating the development of receptors for ___ secreted by other cells.

10. Malignant tumors are characterized by the ability to spread, or ___.

11. Posttranslational modification for export proteins takes place in the___.

12. The ___ is the genetic makeup of a population.

13. A new DNA helix is synthesized in short segments that must be put together end-to-end by an enzyme called ___.

14. The law of ___ enables one to predict the nucleotide sequence of one DNA strand when the sequence of the other is known.

15. Cytosine, uracil, and thymine are single-ringed nitrogenous bases called ___.

B. Matching

A. secretory vesicle   G. transport vesicle   M. anticodon   S. nucleotide
B. chromatin   H. G_2 phase   N. exon   T. metastasis
C. somatic cells   I. regulatory proteins   O. S phase   U. base triplet
D. proto-oncogenes   J. nucleolus   P. codon   V. genetic code
E. translation   K. nucleosomes   Q. genotype   W. transcription
F. metaphase   L. kinetochore   R. lysosome   X. germ cells

1. Appears divided into nucleosome segments

2. A Golgi vesicle destined to release a cell product by exocytosis

3. Contains protein that causes movement of the chromosomes during cell division

4. System that allows four nucleotides to code for amino acid sequence of all proteins

5. The only haploid cells in the body

6. A sugar, a phosphate, and a nitrogenous base

7. Contain 23 pairs of homologous chromosomes in *Homo sapiens*

8. Process during which polyribosomes are active

9. Process in which mRNA is made from DNA in the nucleus

10. Turn genes on or off; are activated by chemical messengers

C. True or False

1. Some codons do not code for amino acids.

2. In DNA, a pyrimidine always pairs with a purine.

3. DNA cannot leave the cell nucleus to participate in cytoplasmic protein synthesis.

4. Proteins can only be synthesized on the rough endoplasmic reticulum.

5. The job of tRNA is to read the anticodons on mRNA.

6. A person's phenotype depends mainly on his/her genotype but can be influenced by the environment.

7. A codon is a segment of DNA that codes for one specific polypeptide.

8. Sex-linked traits can be inherited only by males.

9. Ribosomal RNA contains no thymine.

10. Recessive alleles are much less common in the population than are dominant alleles.

D. Multiple Choice

1. Four of the following are nitrogenous bases in DNA or mRNA, while ___ is an amino acid.
   (a) glycine; (b) adenine; (c) cytosine; (d) thymine; (e) uracil

2. Somatic cells contain ___ chromosomes in humans. (a) a haploid set of; (b) a diploid set of; (c) 23; (d) 44; (e) X and Y

3. Colon cancer results from mutations in three or more: (a) proto-oncogenes; (b) oncogenes; (c) tumor-suppressor genes; (d) growth factors; (e) carcinogens.

4. Errors in DNA replication are usually detected and corrected by a proofreader molecule of: (a) DNA helicase; (b) reverse transcriptase; (c) recombinant DNA; (d) DNA polymerase; (e) DNA ligase.

5. The difference between pre-mRNA and mRNA is that mRNA lacks noncoding segments called: (a) introns; (b) exons; (c) axons; (d) bosons; (e) chevrons.

6. All carcinogens are ___, but not all of these are necessarily carcinogens. (a) mutagens; (b) teratogens; (c) plasmagens; (d) glycogens; (e) antigens

7. The period in which a cell is preparing to divide is: (a) the S phase; (b) the G_1 phase; (c) the G_2 phase; (d) the M phase; (e) telophase.

8. When assembling a protein, a ribosome reads the "instructions" on mRNA and is able to select the correct tRNA by recognizing its: (a) tertiary conformation; (b) codon; (c) anticodon; (d) uracil; (e) TATA box.

9. DNA wound around a cluster of eight proteins forms a granule in the nucleus called a: (a) nucleosome; (b) histone; (c) ribosome; (d) histosome; (e) chaperone.

10. All are true about the Human Genome Project and its results *except:* (a) it was an international effort; (b) it demonstrated that there are fewer genes than previously believed; (c) it showed that one protein can code for more than one gene; (d) members of our species are over 99% genetically identical; (e) it increased our knowledge of disease-causing mutations and their loci.

11. ABO blood groups is a trait that demonstrates ___ inheritance. (a) multiple allele; (b) polygenic; (c) pleiotropy; (d) sex-linked; (e) incomplete dominant

12. Males are more likely than females to inherit a genetic disease if the gene for it is: (a) dominant; (b) pleiotropic; (c) X-linked; (d) mutated; (e) polygenic.

13. All of these are true about the comparisons between DNA and RNA *except:* (a) DNA contains the bases G, C, A, and T, while RNA contains U instead of T; (b) DNA is a very large molecule compared to RNA; (c) both types of nucleic acids tend to be formed into a double helix; (d) both contain phosphate-and-sugar backbones; (e) DNA never leaves the nucleus, while RNA carries the genetic code to cytoplasmic ribosomes for "reading."

14. UAG, UGA, and UAA are: (a) stop codons on mRNA; (b) anticodons on mRNA that code for amino acids; (c) nonsense code words on DNA; (d) codons for glycine, alanine, and valine, respectively; (e) 3 of 20 code words essential in transcription.

15. The observable traits of an individual that result from his/her genes are called the: (a) genotype; (b) karyotype; (c) archetype; (d) phenotype; (e) homotype.

16. DNA contains all of the following *except:* (a) deoxyribose; (b) uracil; (c) phosphorus; (d) adenine; (e) cytosine.

17. Which of the following base pairs is *not* normal for DNA or mRNA? (a) T–A; (b) G–C; (c) A–U; (d) A–G; (e) C–G

18. All coding and noncoding DNA contained in an individual's haploid (n) set of chromosomes is his/her: (a) gene pool; (b) genealogy; (c) nucleotides; (d) genome; (e) phenotype.

19. A person with blood type alleles I^A and I^B shows characteristics of both blood types A and B because these alleles are: (a) pleiotropic; (b) both recessive; (c) codominant; (d) homozygous; (e) polygenic.

20. A Punnett square is used to: (a) show an individual's chromosomes in pairs and in order by size; (b) find the locus of a particular gene on a chromosome; (c) distinguish dominant alleles from recessive alleles; (d) show how haploid cells become diploid; (e) show the expected outcome of a single mating.

21. Which of these is/are true?
    1. Watson and Crick published their findings on the structure of DNA without performing any experiments.
    2. Rosalind Franklin's X-ray diffraction work contributed to unraveling the mystery of DNA structure.
    3. Gregor Mendel observed the patterns of heredity in peas which later led to a better understanding of inheritance patterns.
    4. Wilkins, Franklin, and Watson and Crick received the Nobel Prize for their work on DNA.

    (a) 1 & 3; (b) 2 & 4; (c) 1, 2, & 3; (d) 4 only; (e) all the above

22. Genes:
    1. are segments of DNA that code for a protein or a group of proteins.
    2. determine the unique characteristics of a species.
    3. are hereditary units passed from one generation to the next.
    4. are blue denim garments advertised by anorexic models.

    (a) 1 & 3; (b) 2 & 4; (c) 1, 2, & 3; (d) 4 only; (e) all the above

23. Which of these is/are true in DNA replication?
    1. Helicase produces replication forks.
    2. New histones are imported into the nucleus after replication is complete, so that new nucleosomes can be formed.
    3. DNA polymerase moves along each DNA strand, matching exposed bases according to the base pairing rule.
    4. At the completion of semiconservative replication, two molecules of DNA made from free nucleotides are formed.

    (a) 1 & 3; (b) 2 & 4; (c) 1, 2, & 3; (d) 4 only; (e) all the above

24. Which of these is/are true about genetics?
    1. Two homologous chromosomes look alike because they are genetically identical.
    2. Recessive alleles can be expressed in a child only if both parents are carriers.
    3. Males tend to have dominant genes, while females tend to have recessive alleles.
    4. Different forms of a gene at the same locus on homologous chromosomes are called alleles.

    (a) 1 & 3; (b) 2 & 4; (c) 1, 2, & 3; (d) 4 only; (e) all the above

25. Which of these is/are true?
    1. Barring mutations, most somatic cells in the body have identical genes.
    2. The backbone of the DNA molecule is made of alternating pyrimidine and purine base pairs.
    3. Mutations in signal peptides can cause disease.
    4. In the synthesis of protein "X," 106 amino acids are used; therefore, 53 ATP molecules are used in making "X."

    (a) 1 & 3; (b) 2 & 4; (c) 1, 2, & 3; (d) 4 only; (e) all the above

E. Word Origins

1. In *chromosome, chromo-* means "genes."
2. In *mutation, muta-* means "deformed."
3. In *angiogenesis, angio-* means "vessel."
4. In *metaphase, meta-* means "next."
5. In *telophase, telo-* means "end."
6. In *kinetochore, kineto-* means "central."
7. In *neoplasm, neo-* means "cancer."
8. In *malignant, mal-* means "cancer."
9. In *mutagen, -gen* means "producing."
10. In *haploid, hapl-* means "less."
11. hetero-
12. -morph-
13. onco-
14. ana-
15. -dacty

F. Which One Does Not Belong?

1. (a) UAG; (b) GAG; (c) GCA; (d) CAU

2. (a) tissue growth; (b) production of sex cells; (c) replacement of old cells; (d) repair of injured tissues

3. (a) codominance; (b) pleiotropy; (c) polygenic inheritance; (d) heterozygous

4. (a) herpes simplex type 2; (b) UV radiation; (c) peroxisome enzymes; (d) cigarette tar

5. (a) cancer; (b) malignant; (c) metastatic; (d) benign

G. Genetics Application Problem

Susan and Bill both have normal color vision, but Susan's father was colorblind. Jennifer, their first child, was two when they discovered she was colorblind. Bill sued for divorce on grounds of infidelity. Without doing any expensive genetic testing, did Bill have a case? Why or why not? If the child had been a colorblind boy, would he have had a case? Why or why not?

5   Histology

A. Short Answer

1. Mature cartilage cells are called ___ *Chondrocyte* .

2. The brain, spinal cord, and nerves arise from an embryonic germ layer called *Ectoderm* ___.

3. Which type of muscle is not striated?   *Smooth*

4. Which type of muscle has multinucleated cells?   *skeletal*

5. Name a connective tissue that does not usually contain fibers. *blood*

6. Nearly every epithelium rests on a layer of ___ *Areola* tissue. (Be specific.)

7.   Cartilage that covers the end of a bone at a movable joint is called ___ *hyaline* cartilage.

8. *Macrophages* ___ are large phagocytes that inhabit fibrous connective tissues where they remove harmful bacteria.

9. Fat is also called *Adipose* ___ tissue, and its cells are called ___ *Adipocytes* .

10. Most endocrine glands secrete their products into the ___ *blood*

11. A gland whose secretion consists of disintegrated cells is called a(n) ___ *Holocrine* gland.

12. The shrinkage of a tissue due to aging is called ___ *Senile atrophy* . [*two words*]

13. *Goblets* ___ cells in simple columnar and pseudostratified columnar epithelium secrete mucus.

14. The fluid between cells of a tissue is called ___ *Interstitial fluid*

15. In the process of ___ *Tissue Engineering* , artificial tissues and organs are produced in a laboratory for use in treating some birth defects and disfiguring injuries.

B. Matching

A. transitional   G. tight junctions   M. cardiac   S. skeletal
B. collagen   H. adipocytes   N. metaplasia   T. mesoderm
C. apoptosis   I. ectoderm   O. histiocytes   U. lamina propria
D. hyaluronic acid   J. striations   P. intercalated discs   V. desmosomes
E. gap junctions   K. dense irregular   Q. endoderm   W. chondrocytes
F. stem cells   L. smooth   R. necrosis   X. connective

1. Most abundant and histologically variable tissue type

2. The only type of muscle under voluntary control

3. The connective tissue immediately deep to the epithelium of a mucous membrane

4. Muscle type containing branched, uninucleate cells

5. Embryonic origin of mucous membranes in the respiratory tract

6. Comprises 25% of the body's protein

7. Cell type that retains developmental plasticity

8. Change from one type of mature tissue to another

9. Intercellular junctions that allow ions and small molecules to pass from the cytoplasm of one cell into the next

10. Pathological tissue death

C. True or False

1. Transitional epithelium is found only in the urinary and respiratory tracts.

2. In a pseudostratified epithelium, all cells touch the basement membrane.

3. Most embryonic stem cells come from aborted fetuses.

4. Both epithelia and cartilage are avascular tissues.

5. Columnar epithelium is found in the heart, among other places.

6. Both muscle and bone arise from embryonic mesoderm.

7. Since cuboidal cells are thicker than squamous cells, stratified cuboidal epithelium is stronger than stratified squamous epithelium.

8. The peritoneal cavity is lined by a moist mucous membrane.

9. There are only four primary tissue types in the human body.

10. Apoptosis is programmed cell death.

D. Multiple Choice

1. Nervous tissue consists of two cell types, neurons and the more abundant: (a) fibroblasts; (b) axons; (c) glia; (d) chondrocytes; (e) osteocytes.

2. Intercellular junctions called intercalated discs are an aid to identifying: (a) areolar tissue; (b) osseous tissue; (c) cardiac muscle; (d) smooth muscle; (e) nervous tissue.

3. Functions of epithelial tissue include all *except:* (a) vascularization; (b) excretion; (c) sensation; (d) protection; (e) filtration.

4. Which of these is not a connective tissue? (a) blood; (b) muscle; (c) cartilage; (d) areolar tissue; (e) osseous tissue

5. Mast cells and fibroblasts are characteristic of: (a) the epithelia of mucous membranes; (b) all types of cartilage; (c) hyaline cartilage only; (d) mastoid tissue; (e) areolar tissue.

6. The components of a connective tissue are cells, fibers, and: (a) a basement membrane; (b) cytosol; (c) protoplasm; (d) ground substance; (e) stroma.

7. Elastic cartilage can be found in: (a) the external ear; (b) tracheal rings; (c) the walls of blood vessels; (d) the wall of the esophagus; (e) tendons and ligaments.

8. The most likely place to find adipocytes is in: (a) the hypodermis; (b) osseous tissue; (c) dense regular connective tissue; (d) hyaline cartilage; (e) mucous membranes.

9. The least vascular of the following tissues is: (a) cardiac muscle; (b) epithelium; (c) bone; (d) areolar tissue; (e) adipose tissue.

10. A tissue specialized for energy storage and thermal insulation is: (a) cartilage; (b) muscular tissue; (c) adipose tissue; (d) epithelium; (e) nervous tissue.

11. A decrease in the size of a tissue or organ is: (a) hyperplasia; (b) metaplasia; (c) hypertrophy; (d) atrophy; (e) apoptosis.

12. The type of epithelium that is most commonly ciliated is: (a) simple cuboidal; (b) simple squamous; (c) pseudostratified; (d) transitional; (e) stratified squamous.

13. The epithelium best suited for resisting abrasion is: (a) pseudostratified; (b) transitional; (c) simple squamous; (d) simple columnar; (e) stratified squamous.

14. In a gland, the cells that synthesize the secretion and conduct it to the surface constitute the: (a) parenchyma; (b) stroma; (c) lamina propria; (d) integument; (e) endothelium.

15. The formation of scar tissue is called: (a) regeneration; (b) necrosis; (c) neoplasia; (d) metaplasia; (e) fibrosis.

16. The only type of cartilage that never has a perichondrium is: (a) collagenous; (b) osseous; (c) fibrocartilage; (d) hyaline; (e) elastic.

17. Adult human organs are made of all of the following *except:* (a) fibrous; (b) epithelial; (c) nervous; (d) muscular; (e) connective.

18. Heat production is a function of both ___ tissue. (a) nervous and muscular; (b) adipose and muscular; (c) epithelial and connective; (d) nervous and connective; (e) osseous and reticular

19. ___ tissue has an abundance of open, fluid-filled space, allowing for rapid migration of immune cells and defense against pathogens. (a) Dense irregular; (b) Adipose; (c) Nervous; (d) Hyaline; (e) Areolar

20. Tendons and ligaments are made of: (a) muscular tissue; (b) dense irregular connective tissue; (c) osseous tissue; (d) areolar tissue; (e) dense regular connective tissue.

21. Connective tissue:
    1. includes fewer cells than does epithelial tissue.
    2. binds organs together.
    3. is made up of large areas of ground substance containing glycosaminoglycan.
    4. varies in vascularity.

    (a) 1 & 3; (b) 2 & 4; (c) 1, 2, & 3; (d) 4 only; (e) all the above

22. Which of these is/are true about cartilage?
    1. It tends to be more vascular than other connective tissues.
    2. The tip of the nose and external ear are made of hyaline cartilage.
    3. It is classified by the types of chondrocytes it contains.
    4. Chondroblasts become trapped in lacunae as they secrete matrix around themselves.

    (a) 1 & 3; (b) 2 & 4; (c) 1, 2, & 3; (d) 4 only; (e) all the above

23. Which of these is/are true?
    1. GAGs help regulate water and electrolyte balance in tissues.
    2. Mast cells secrete heparin and antihistamine.
    3. Proteoglycans help prevent the spread of disease-causing organisms.
    4. The most abundant type of protein in the human body is hemoglobin.

    (a) 1 & 3; (b) 2 & 4; (c) 1, 2, & 3; (d) 4 only; (e) all the above

24. Which of these is/are true?
    1. Epithelia are capable of metaplasia.
    2. Growth by increasing cell number is called hyperplasia.
    3. Pemphigus vulgaris causes desmosomes to break down, leading to tissue fluid loss.
    4. Ground substance in blood consists of blood proteins secreted by erythrocytes.

    (a) 1 & 3; (b) 2 & 4; (c) 1, 2, & 3; (d) 4 only; (e) all the above

25. Which of these is/are true?
    1. Apoptosis occurs throughout life.
    2. Marfan's syndrome is caused by a mutated gene that codes for a protein found in connective tissue.
    3. Nerve and muscle tissue exhibit membrane potentials.
    4. Platelets are cell fragments that function in blood clotting.

    (a) 1 & 3; (b) 2 & 4; (c) 1, 2, & 3; (d) 4 only; (e) all the above

E. Word Origins

1. In *ectoderm, -derm* means "skin."
2. In *mesoderm, meso-* means "under."
3. In *reticular, reti-* means "striped."
4. In *fibroblast, -blast* means "destroy."
5. In *reticular, -icul* means "little."
6. In *leukocyte, leuko-* means "blood."
7. In *chondrocyte, chondro-* means "cartilage."
8. In *periosteum, peri-* means "near."
9. In *desmosome, desmo-* means "basal."
10. In *merocrine, -crin* means "secrete."
11. inter-
12. -trophy
13. apo-
14. hyal-
15. kerat-

F. Which One Does Not Belong?

1. (a) fibroblast; (b) mast cell; (c) chondroblast; (d) adipocyte

2. (a) areolar; (b) dense regular; (c) adipose; (d) reticular

3. (a) cartilage; (b) blood; (c) bone; (d) nerve

4. (a) blood; (b) muscle; (c) nerve; (d) epithelium

5. (a) cutaneous; (b) mucous; (c) serous; (d) synovial

G. Matching Exercise

Match the tissue with its function and/or structure.

A.   simple squamous
B.   simple cuboidal
C.   simple columnar
D.   pseudostratified columnar
E.   keratinized stratified squamous
F.   nonkeratinized stratified squamous
G.   stratified cuboidal
H.   transitional epithelium
I.   areolar tissue
J.   reticular tissue
K.   adipose tissue

L. dense regular
M. dense irregular
N. hyaline cartilage
O. elastic cartilage
P. fibrocartilage
Q. bone
R. blood
S. nerve
T. skeletal muscle
U. cardiac muscle
V. smooth muscle

1.   Most attached to bones; body movement, facial expression, breathing, speech
2.   Found on bone ends, eases joint movements, keeps airways open
3.   Contained in heart and blood vessels; gas transport, defense
4.   Stretches to allow filling of bladder, ureter
5.   Found only on palms and soles
6.   Underlies most epithelia, loosely binds them to deep tissues; surrounds nerves, vessels
7.   Resists compression and absorbs shock in intervertebral discs
8.   Layers of flat cells that resist pathogens in oral cavity, vagina
9.   Absorption and secretion, lining of intestine, uterus
10. Allows rapid diffusion, secretes serous fluid; found in alveoli and lining of heart
11. Durable, withstands multiple stresses; found in deep dermis and capsules around viscera
12. Binds bones together, attaches muscle to bone
13. Upper respiratory tract; secretes and moves mucus
14. Absorption and secretion, produces mucous coat; found in thyroid follicles
15. Supportive stroma for lymph nodes, spleen, thymus

6   The Integumentary System

A. Short Answer

1. Thin skin is a type that has relatively little stratum _corneum_

2. The majority of cells in the epidermis are a type called _Keratinocyte_

3. An epidermal cell involved in the sense of touch is called a(n) _tactile_ cell.

4. Dead cells of the epidermal surface are said to _desquamate_ as they flake off.

5. The epidermis cannot slide freely over the dermis because of projections of dermis similar to corrugated cardboard, called ___. _dermal papillae_

6. A bruise is also called a(n) _hematoma_

7. The action of ultraviolet light on dehydrocholesterol circulating through the skin is the first step in the synthesis of ___. _Vitamin D_

8. When it is cold, cutaneous _vasoconstriction_ helps retain heat by reducing blood flow through the skin.

9. Coarse, pigmented hairs are called _Terminal_ hairs.

10. The middle of a hair, called the _medulla_, is composed of loosely arranged cells and air spaces.

11. _debridement_ is the removal of eschar from burned skin.

12. The growth zone of a fingernail or toenail is called the _Nail matrix_

13. The oily secretion of the scalp comes from the _sebaceous_ glands associated with the hair follicles.

14. A(n) _Second degree_ burn destroys epidermal and some dermal tissue but leaves some of the dermis intact.

15. Even though it accounts for only 5% of skin cancer, _melanoma_ is the most deadly form.

B. Matching

A. hemangioma	G. exfoliate	M. medulla	S. thermoregulation
B. jaundice	H. papillary layer	N. acid mantle	T. basal cell carcinoma
C. reticular layer	I. hematoma	O. arrector pili	U. apocrine glands
D. stratum basale	J. merocrine glands	P. nevus	V. squamous cell carcinoma
E. keratinocytes	K. melanoma	Q. piloerection	W. stratum corneum
F. hypodermis	L. bulb	R. cyanosis	X. subcutaneous muscle

1. Cells that contain most of the melanin in black or brown skin *Keratinocyte*

2. The more superficial of the two layers of dermis *papillary layer*

3. Blueness of the skin due to an oxygen deficiency *Cyanosis*

4. Dense irregular connective tissue in deep layers of dermis *reticular layer*

5. A protective chemical film that inhibits bacterial growth on the skin *Acid mantle*

6. A homeostatic effect of cutaneous vasoconstriction and vasodilation *thermoregulation*

7. The dilated base of a hair containing a dermal papilla with blood vessels *bulb*

8. Muscle that causes a hair to stand on end when it is cold *arrector pili*

9. Raised pigmented area of skin, sometimes hairy *Nevus*

10. Location of most subcutaneous fat *hypodermis*

C. True or False

1. Basal and squamous cell carcinomas are benign forms of skin cancer. *F*

2. All new epidermal cells are produced by the stratum basale. *T*

3. Death from a severe burn results from fluid loss, infection, and eschar. *T*

4. Jaundice is a yellowish skin color caused by the pigment carotene. *F*

5. The dermis is made of stratified squamous epithelium. *F*

6. The most abundant protein of the epidermis is keratin, while the most abundant protein of the dermis is collagen. *T*

7. Fingerprints augment one's sense of touch. *T*

8. The body hair of children is mostly lanugo. *F*

9. All hair growth results from mitosis in the hair cortex. *F*

10. Hair is composed primarily of keratin. *T*

D. Multiple Choice

1. In the ___, the epidermal cells shrink and pull apart during histological fixation, making their desmosomes more apparent. (a) stratum corneum; (b) stratum lucidum; (c) stratum granulosum; (d) stratum spinosum; (e) stratum basale

2. Each hair receives its blood supply from the: (a) root sheath; (b) arrector pili; (c) hair papilla; (d) medulla; (e) cortex.

3. Which of the following skin discolorations, if seen repeatedly in pediatric examinations of the same patient, would most likely suggest child abuse? (a) sunburn; (b) jaundice; (c) hematoma; (d) hemangioma; (e) pallor

4. The type of gland most associated with hair follicles is a(n): (a) mucous gland; (b) parotid gland; (c) apocrine gland; (d) ceruminous gland; (e) sebaceous gland.

5. The epidermis has a ___ only in palmar and plantar skin. (a) stratum corneum; (b) stratum lucidum; (c) stratum granulosum; (d) stratum spinosum; (e) papillary layer

6. The thickest layer of the epidermis is the: (a) reticular layer; (b) stratum lucidum; (c) stratum granulosum; (d) stratum spinosum; (e) papillary layer.

7. The most urgent issue in treating a patient with extensive third-degree burns is: (a) fluid balance; (b) infection control; (c) pain; (d) toxicity; (e) skin grafting.

8. The American Cancer Society's rule of thumb for recognizing a melanoma is based on four key signs. One of these is: (a) desquamation; (b) a round border; (c) asymmetry with scalloped edges; (d) diameter of the lesion; (e) the discharge of pus.

9. The ___ glands function in evaporative cooling. (a) apocrine; (b) merocrine; (c) holocrine; (d) sebaceous; (e) ceruminous

10. The protein of the stratum corneum is: (a) soft keratin; (b) hard keratin; (c) carotene; (d) collagen; (e) elastin.

11. The pigment of brown-to-black skin is synthesized by: (a) Merkel cells; (b) chromocytes; (c) melanocytes; (d) keratinocytes; (e) mast cells.

12. A pigment that comes from the diet rather than being synthesized in the body is: (a) trichosiderin; (b) hematin; (c) carotene; (d) keratin; (e) melanin.

13. ___ is a type of hair found almost exclusively in fetuses. (a) Lanugo; (b) Terminal hair; (c) Pilus; (d) Nevus; (e) Vellus

14. Pattern baldness is more common in men than in women because: (a) the allele for this trait requires high testosterone levels for its expression; (b) it is an X-linked recessive trait; (c) it is an autosomal trait; (d) it is suppressed by estrogen in women; (e) only men secrete testosterone.

15. The ___ glands of humans are probably comparable to the scent glands of other animals in their function. (a) merocrine; (b) ceruminous; (c) sebaceous; (d) mammary; (e) apocrine

16. Which of these is *not* true about UVR? (a) decreases with elevation; (b) stimulates vitamin D synthesis in keratinocytes; (c) can cause infertility; (d) is one selection pressure for variations in skin pigmentation; (e) accounts for over 75% of skin color variation in humans

17. Unusual redness of the skin is called: (a) cyanosis; (b) erythema; (c) jaundice; (d) melanoma; (e) pallor.

18. Which of these is *not* considered an appendage of the skin? (a) a fingernail; (b) a hair; (c) a dermal papilla; (d) an apocrine gland; (e) a mammary gland

19. The color of blonde hair results from: (a) trichosiderin; (b) air; (c) light refraction; (d) melanin; (e) carotene.

20. The ___ of a toenail has the same function as the stratum basale of a hair follicle. (a) eponychium; (b) hyponychium; (c) body; (d) bed; (e) matrix

21. Which of these is/are true?
    1. Mammary glands are modified apocrine sweat glands.
    2. People who use sunscreen have a greater chance of getting some forms of skin cancer than those who don't use it.
    3. Cerumen keeps the eardrum pliable.
    4. Hemangiomas are particularly malignant forms of skin cancer.

    (a) 1 & 3; (b) 2 & 4; (c) 1, 2, & 3; (d) 4 only; (e) all the above

22. Which of these is/are true?
    1. Dendritic cells originate in bone marrow.
    2. Artificial skin can be used to temporarily cover burns.
    3. The dermis contains temperature sensors.
    4. Diaphoresis is insensible perspiration.

    (a) 1 & 3; (b) 2 & 4; (c) 1, 2, & 3; (d) 4 only; (e) all the above

23. Which of these is/are true?
    1. Next to muscle, the skin is the largest organ of the body.
    2. Thick skin contains all the same glands associated with thin skin.
    3. *Dermatophagoides* feeds on house dust.
    4. The number of melanocytes is about the same in people of all races.

    (a) 1 & 3; (b) 2 & 4; (c) 1, 2, & 3; (d) 4 only; (e) all the above

24. Which of these is/are true about hair?
    1. Alopecia is the opposite of hirsutism.
    2. People with albinism lack melanocytes.
    3. Eumelanin causes brown or black hair color, while pheomelanin causes red hair.
    4. Wavy hair is round in cross section.

    (a) 1 & 3; (b) 2 & 4; (c) 1, 2, & 3; (d) 4 only; (e) all the above

25. Skin color can be determined by the pigment(s):
    1. erethema.
    2. carotene.
    3. jaundice.
    4. melanin.

    (a) 1 & 3; (b) 2 & 4; (c) 1, 2, & 3; (d) 4 only; (e) all the above

E. Word Origins

1. In *epidermis, epi-* means "upon, above."
2. In *keratinocyte, kera-* means "tough."
3. In *papilla, papi* means "nipple."
4. In *lucidum, lucid-* means "clear."
5. In *linea alba, linea* means "line."
6. In *hemangioma, angi-* means "mass, tangle."
7. In *sudoriferous, sudor-* means "bad odor."
8. In *sudoriferous, fer-* means "bear, carry."
9. In *diaphoresis, dia-* means "through."
10. In *hyponychium, onychi-* means "cuticle."
11. dendr-
12. melan-
13. -oma
14. cata-
15. hirsut-

F. Which One Does Not Belong?

1. (a) dendritic cells; (b) keratinocytes; (c) tactile cells; (d) melanocytes

2. (a) nail matrix; (b) stratum granulosum; (c) hair matrix; (d) stratum basale

3. (a) apocrine; (b) sudoriferous; (c) merocrine; (d) ceruminous

4. (a) lunule; (b) eponychium; (c) corneum; (d) hyponychium

5. (a) sunburn prevention; (b) thermoregulation; (c) dehydration prevention; (d) vitamin D synthesis

G. Figure Exercise

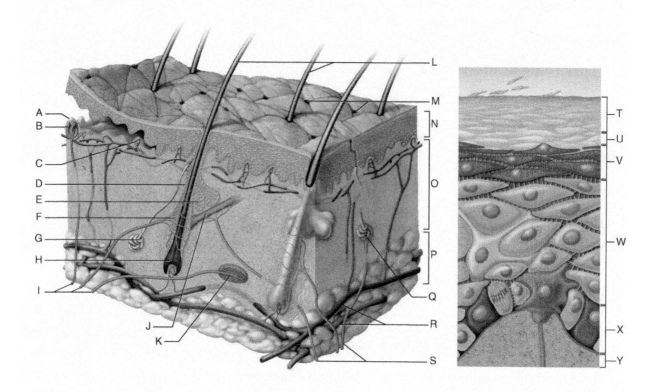

Match the statements with the structure in the diagrams.

1. Produces the most sweat when we exercise
2. Gives off dander
3. Produces glycolipids that waterproof the skin
4. Contains melanocytes, tactile cells, and keratinocytes
5. Holocrine glands that secrete contents of broken-down cells
6. Causes hair to move in skin in response to stimuli
7. Region containing most glands
8. Especially abundant in the groin and axillary regions
9. Present only in thick skin where keratinocytes are densely packed with eleidin
10. Home of dendritic cells that protect against infections

7   Bone Tissue

A.   Short Answer

1. Osteoblasts secrete the hormone ___, that stimulates insulin secretion by the pancreas. *[handwritten: Osteocalcin]*

2. The skull, spinal column, and thoracic cage make up the ___ division of the skeleton. *[handwritten: Axial]*

3. Osteocytes communicate with each other through their slender cytoplasmic extensions that lie in *[handwritten: Canaliculi]*

4. Histological examination reveals two types of bone: ___ bone, which forms the surface of all bones, and ___ bone, found within the bodies of the vertebrae and heads of the long bones, for example.

5. In life, a bone is covered with a very tough, fibrous sheath called the ___. *[handwritten: Periosteum]*

6. The bone-forming cells that synthesize its matrix and later become trapped in it are called ___. *[handwritten: Osteoblast]*

7. Most lamellae of compact bone are concentrically arranged around a space called the ___ *[handwritten: Central Canal]*

8. Blood is produced in an organ of the skeletal system called ___ *[handwritten: RBM (myeloid tissue]*

9. ___, which digests collagen in bone matrix, is secreted by osteoclasts.

10. Lack of mineral deposition causes bone softening, a condition known as ___, while lack of protein causes ___ in which bone becomes excessively brittle.

11. The growth zone of a child's femur is a cartilaginous region between the diaphysis and epiphysis of the bone called the ___.

12. According to ___, the structure of a bone is determined by the <u>mechanical</u> and <u>gravitational</u> forces applied to that bone. *[handwritten: Wolff law]*

13. A deficiency of calcium in the blood, called ___, can lead to sustained muscle contractions called ___ *[handwritten: hypocalcemia; tetany]*

14. Calcium homeostasis in adults is maintained almost entirely by the secretion of ___, which regulates the activity of bone-dissolving cells called ___ *[handwritten: parathyroid; Osteoclast]*

15. The most common adult bone disease, especially among postmenopausal white women, is ___ *[handwritten: Osteoporosis]*

B. Matching

A. intramembranous	G. Wolff's law	M. osteoclasts	S. blood vessels
B. medullary cavity	H. osteocytes	N. periosteum	T. monocytes
C. diaphysis	I. sesamoid bones	O. central canals	U. articular cartilage
D. lymphocytes	J. red blood cells	P. symphysis	V. endochondral
E. chondrocytes	K. osteogenic cells	Q. sutural	W. wormian bones
F. osteoblasts	L. epiphyses	R. hemopoiesis	X. appositional

1. Occupy the lacunae of compact bone

2. Bones that develop within the tendons sometime after birth

3. Region of long bone made of compact bone with central medullary cavity

4. Type of ossification that produces the flat bones of the cranium

5. Bone-dissolving macrophages with ruffled borders

6. Cells that release RANKL that in turn increases the number of osteoclasts

7. Contain the nearest blood vessels to most osteocytes in compact bone

8. Describes the function of red bone marrow

9. Mode of ossification that produces the flat cranial bones

10. The only method by which mature bones can grow

C. True or False

1. Without medical care, a person could not live for more than a few days without the parathyroid glands.

2. Bones attain their final mass, size, and shape at the end of adolescence.

3. Most bones are in the appendicular skeleton.

4. The epiphyseal plate is the primary site of bone growth in people of all ages.

5. Spongy bone consists of a random array of calcified trabeculae, like the fibers of a kitchen sponge.

6. Without calcitonin, most of us would develop fatal hypercalcemia.

7. Calcium is not deposited in osseous tissue unless there is also a proportionate amount of phosphate in the blood.

8. Some bones are formed by both endochondral and intramembranous ossification.

9. A deficiency of dietary vitamin D will almost certainly lead to hypocalcemia.

10. Yellow bone marrow can change into hemopoietic tissue when necessary.

D. Multiple Choice

1. Hypocalcemia causes: (a) diarrhea; (b) sluggishness; (c) changes in blood phosphate concentrations; (d) hypercalcemia; (e) excessive sensitivity in nervous and muscle tissues.

2. During ossification of the humerus, ___ hypertrophy and die. (a) osteocytes; (b) osteogenic cells; (c) osteoblasts; (d) chondrocytes; (e) chondroblasts

3. All of these are true about calcitonin *except*: (a) it is active mainly in children; (b) it inhibits osteoclast activity; (c) it raises calcium concentration in the blood; (d) it stimulates bone synthesis; (e) it is produced in thyroid C cells.

4. Calcitriol (vitamin D) synthesis involves all of the following *except:* (a) ultraviolet light; (b) cholesterol; (c) the kidneys; (d) the liver; (e) osteoblasts.

5. The skeletal system serves all of the following purposes *except:* (a) acid–base homeostasis; (b) electrolyte homeostasis; (c) blood formation; (d) thermoregulation; (e) protective enclosure of soft organs.

6. Parathyroid hormone (PTH) functions in all of the following *except:* (a) inhibits osteoclasts in children; (b) indirectly stimulates RANKL's release; (c) stimulates calcium reabsorption from the kidney; (d) aids in the formation of calcitriol; (e) increases blood calcium levels.

7. The process in which hyaline cartilage is converted to bone is called ___ ossification. (a) interstitial; (b) epiphyseal; (c) intramembranous; (d) appositional; (e) endochondral

8. Calcitriol functions in: (a) osteoclast inhibition; (b) release of osteoclast-inhibiting factor; (c) prevention of bone breakdown; (d) inhibition of collagen deposition by osteoblasts; (e) increase of calcium absorption by the small intestine.

9. One example of ectopic ossification is: (a) osteoporosis; (b) compact bone; (c) tetany; (d) atherosclerosis; (e) osteogenic sarcoma.

10. A pit in the surface of a bone is called a: (a) trochanter; (b) condyle; (c) tubercle; (d) fovea; (e) fissure.

11. A rough area on a bone, usually serving for muscle attachment, is a(n): (a) fossa; (b) tuberosity; (c) alveolus; (d) facet; (e) sulcus.

12. The diaphysis of a 35-year-old's femur is normally filled with: (a) yellow marrow; (b) red marrow; (c) diploe; (d) endosteum; (e) spongy bone.

13. Tetany, laryngospasm, and suffocation could result from: (a) osteoporosis; (b) lack of parathyroid hormone; (c) lack of calcitonin; (d) hypercalcemia; (e) osteomalacia.

14. At one stage a healing fracture exhibits a(n) ___, containing collagen and patches of fibrocartilage. (a) soft callus; (b) hard callus; (c) hematoma; (d) chondrosarcoma; (e) osteoma

15. The relative concentrations of calcium and phosphate ions must reach a critical value, the solubility product, in order: (a) to avoid hypocalcemia; (b) for them to be excreted together in the urine; (c) to avoid achondroplastic dwarfism; (d) for bone resorption to occur; (e) for bone deposition to occur.

16. The carpals are bones of the: (a) cranium; (b) feet; (c) ankles; (d) thoracic cage; (e) wrist.

17. The head of a long bone is called the: (a) metaphysis; (b) trabeculae; (c) trochanter; (d) epiphysis; (e) epicondyle.

18. The fusion of bone marrow stem cells gives rise to: (a) chondrocytes; (b) osteocytes; (c) osteoclasts; (d) osteogenic cells; (e) osteoblasts.

19. Blood vessels of the periosteum enter the bone matrix by way of: (a) central canals; (b) perforating canals; (c) canaliculi; (d) fossae; (e) the medullary cavity.

20. In endochondral ossification of the metaphyses of a long bone, the zone of _____ is the area where chondrocytes quit dividing and increase in size. (a) hypertrophy; (b) reserve cartilage; (c) calcification; (d) proliferation; (e) bone deposition

21. Which of these is/are true?
    1. Trabeculae of spongy bone are covered with endosteum.
    2. The inner layer of the periosteum contains osteoblasts.
    3. Perforating (Sharpey's) fibers anchor tendons to bone.
    4. Diploe is hemopoietic in adults.

    (a) 1 & 3; (b) 2 & 4; (c) 1, 2, & 3; (d) 4 only; (e) all the above

22. Which of these is/are true about bone structure?
    1. Bone matrix contains proteins like collagen that provide flexibility.
    2. Bone contains both organic and inorganic components.
    3. In most bones, the epiphyses are wider than the diaphysis, and they are places for muscle attachment.
    4. Most of the inorganic mineral in bone matrix is calcium carbonate.

    (a) 1 & 3; (b) 2 & 4; (c) 1, 2, & 3; (d) 4 only; (e) all the above

23. Which of these is/are true about bone cells?
    1. Osteoblasts deposit the organic components of bone.
    2. Osteoclasts are isolated in lacunae in mature compact bone.
    3. Osteocytes do not produce bone matrix.
    4. Osteoporosis is caused by the presence of too many osteoclasts.

    (a) 1 & 3; (b) 2 & 4; (c) 1, 2, & 3; (d) 4 only; (e) all the above

24. Which of these is/are true about conditions that affect the skeleton?
    1. In achondroplastic dwarfism, the skeleton remains cartilaginous throughout life.
    2. Excessive cola consumption can lead to bone loss in women.
    3. Anabolic steroid use allows male athletes to grow taller but only if they begin use in adolescence.
    4. Osteosarcoma is the most deadly form of bone cancer, and it affects males more than it does females.

    (a) 1 & 3; (b) 2 & 4; (c) 1, 2, & 3; (d) 4 only; (e) all the above

25. Which of these is/are true about osteoporosis?
    1. It is the most common bone disease.
    2. Both males and females suffer from it.
    3. It involves a decrease in bone mass, especially in spongy bone.
    4. Risk factors include gender, age, smoking, poor nutrition, and sedentary lifestyle.

    (a) 1 & 3; (b) 2 & 4; (c) 1, 2, & 3; (d) 4 only; (e) all the above

E. Word Origins

1. In *osteocyte, osteo-* means "bone."
2. In *sesamoid, -oid* means "false."
3. In *diaphysis, -physis* means "shaft."
4. In *osteoclast, -clast* means "destroy."
5. In *hemopoietic, -poietic* means "harmonious."
6. In *ectopic, top-* means "place."
7. In *achondroplastic, a-* means "without."
8. In *hypocalcemia, -emia* means "blood condition."
9. In *hypocalcemia, calc-* means "calcium."
10. In *orthopedics, ortho-* means "skeleton."
11. -blast
12. epi-
13. icul
14. peri-
15. chondr-

F. Which One Does Not Belong?

1. (a) osteocyte; (b) osteoid; (c) osteoclast; (d) osteoblast

2. (a) PTH; (b) calcitonin; (c) hydroxyapatite; (d) calcitriol

3. (a) leg bones; (b) tarsal bones; (c) femur; (d) forearm bones

4. (a) displaced; (b) green stick; (c) comminuted; (d) straight

5. (a) bony collar formation; (b) soft callus formation; (c) hematoma formation; (d) granulation tissue formation

G. Figure Exercise

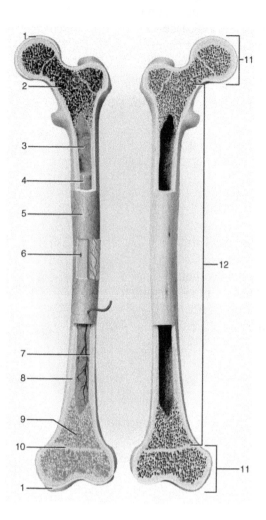

1. ____ is the remnant of the region of growth in length in children.
    (a) 1; (b) 2; (c) 9; (d) 10; (e) 11

2. ____ contains red or yellow marrow, depending on age.
    (a) 3; (b) 4; (c) 6; (d) 8; (e) 9

3. ____ is/are made of hyaline cartilage.
    (a) 1; (b) 1 & 3; (c) 2 & 12; (d) 9; (e) 10

4. ____ is the target for the most common bone disease.
    (a) 1; (b) 3; (c) 8; (d) 9; (e) 10

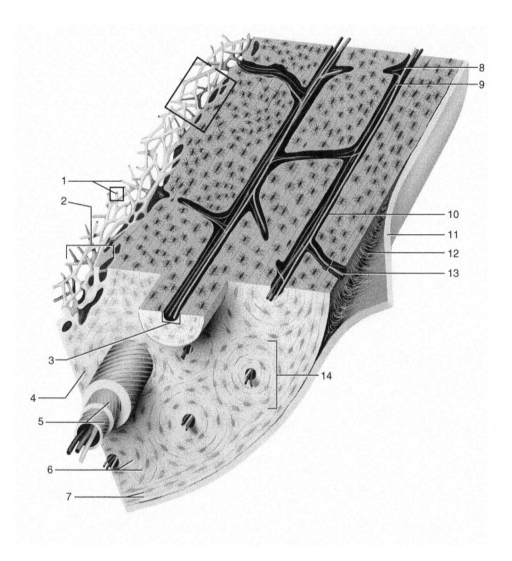

5. Which of these houses living bone cells?
   (a) 3; (b) 4; (c) 5; (d) 9; (e) 12

6. Number 14 is the structural unit of ___ bone.
   (a) spongy; (b) long; (c) compact; (d) epiphyseal; (e) adolescent

7. Number 11 contains:
   (a) osteocytes; (b) blood vessels; (c) osteoblasts; (d) lacunae; (e) perforating canals.

8. Osteogenic cells can be found in:
   (a) 3, 10, & 11; (b) 5 & 6; (c) 11 & 13; (d) 1, 3, & 4; (e) 4, 5, & 6.

9. The structures shown in number 1 are called:
   (a) osteons; (b) trabeculae; (c) perforating fibers; (d) lamellae; (e) lacunae.

10. The function of number 5 is to: (a) increase surface area for hydroxyapatite deposition;
    (b) increase the number of osteocytes in bone; (c) decrease the tension on bone; (d) increase the
    strength of bone; (e) make bone more flexible.

8   The Skeletal System

A. Short Answer

1. The cranial bones are held together by joints called ___.

2. Cranial nerves and blood vessels pass through openings in the skull called ___.

3. The bony palate is composed of two ___ bones and the palatine processes of the ___.

4. The pituitary gland lies in a depression of the ___ bone, a complex cranial bone with greater and lesser wings.

5. The space on the posterior surface of the scapula inferior to the scapular spine is called the ___.

6. The medial bone of the forearm is the ___.

7. The cartilaginous pads between the bodies of adjacent vertebrae are called ___.

8. Spinal nerves pass through gaps between adjacent vertebrae called the ___.

9. At the inferior end of the sternum is a small, pointed bone called the ___.

10. The apex of your shoulder is formed by a plate-like extension of the scapular spine called the ___.

11. The pelvic girdle is made up of the ___.

12. The right and left hip bones are held together anteriorly by a fibrocartilage pad called the ___.

13. When you are sitting, your body weight is supported on the right and left ___, which are thick, rough areas of the hip bones.

14. The bony prominences on the sides of your ankle, just above the top of a dress shoe, are the lateral and medial ___.

15. The bones of the fingers and toes are called the ___.

B. Matching

A. auricular surface   G. ethmoid   M. lambdoid   S. sternum
B. acetabulum   H. atlas   N. coronal   T. hallux
C. pollex   I. humerus   O. ulna   U. temporal
D. axis   J. hip bones   P. lumbar   V. malleoli
E. olecranon   K. acromion   Q. femur   W. mastoid process
F. scapula   L. occipital condyles R. sphenoid   X. trochlear surface

1. Has a unique structure called the dens or odontoid process
2. The point of your elbow where you rest it on a table
3. Bone that contains the stylomastoid foramen
4. Suture named for the Greek letter λ
5. Knobs on an inferior bone of the skull that articulate with the atlas
6. Vertebrae whose superior articular processes face medially
7. Consists of a manubrium, gladiolus, and xiphoid process
8. Its distal end has a capitulum and a trochlea
9. The thumb
10. Contains the acetabulum, greater sciatic notch, and iliac crest

C. True or False

1. The lambdoid suture separates the frontal bone from the parietal bones.

2. The temporalis muscle originates mainly on the temporal bone.

3. The fibula is medial to the tibia.

4. Unlike other cranial foramina, the foramen lacerum is closed in life, and it provides passage for no major nerves or blood vessels.

5. The mandibular notch is between the condylar and coronoid processes.

6. Thoracic vertebrae can be distinguished from others by their costal facets and bifid spinous processes.

7. The zygomatic arch is made up of the zygomatic process of the zygomatic bone and the temporal process of the temporal bone.

8. The capitulum of the humerus articulates with the proximal end of the radial head.

9. Each half of a hip bone consists of three bones in a child but only one bone in an adult.

10. Only the femur has trochanters.

D. Multiple Choice

1. The ___ bone contains numerous air cells, contributes the upper half of the nasal septum, and has pores for the passage of olfactory nerves. (a) frontal; (b) vomer; (c) ethmoid; (d) sphenoid; (e) nasal

2. All of the following can be palpated on a living person *except* the: (a) mastoid process; (b) mental protuberance; (c) suprasternal notch; (d) sella turcica; (e) olecranon.

3. The squamous suture surrounds: (a) a squamous epithelium; (b) the parietal bone; (c) the temporal bone; (d) the sphenoid bone; (e) the ethmoid bone.

4. The fetal skull has a small gap where the frontal, parietal, temporal, and sphenoid bones meet called the: (a) anterior fontanel; (b) posterior fontanel; (c) mastoid fontanel; (d) parietofrontal fontanel; (e) sphenoid fontanel.

5. Where it meets the tibia, the distal end of the femur is covered with: (a) spongy bone; (b) a synovial membrane; (c) a synostosis; (d) the periosteum; (e) articular cartilage.

6. The jelly-like center of an intervertebral disc is called: (a) the gelatinus centralis; (b) the nucleus pulposus; (c) synovial fluid; (d) vitreous humor; (e) tissue gel.

7. Costal facets are found on: (a) the cervical vertebrae; (b) the thoracic vertebrae; (c) all vertebrae; (d) true ribs; (e) all ribs.

8. The spinal column has all of the following curvatures *except* a: (a) cervical curvature; (b) thoracic curvature; (c) lumbar curvature; (d) sacral curvature; (e) pelvic curvature.

9. The linea aspera is unique to the: (a) ulna; (b) atlas; (c) femur; (d) fibula; (e) hip bone.

10. The sesamoid bone embedded in the quadriceps femoris tendon is the: (a) patella; (b) hamate; (c) medial malleolus; (d) parital; (e) navicular.

11. The coracoid process is found on the same bone as the: (a) styloid process; (b) acromion; (c) supra sternal notch; (d) olecranon fossa; (e) gluteal fossa.

12. The talus articulates superiorly with the: (a) scapula; (b) tibia; (c) femur; (d) navicular; (e) fibula.

13. All of the following are bones of the wrist *except* the ___, which is a bone of the ankle. (a) hamate; (b) capitate; (c) trapezoid; (d) cuboid; (e) pisiform

14. Features of the sphenoid bone include all *except* the: (a) foramen rotundum; (b) optic canal; (c) hypoglossal canal; (d) greater wings; (e) pterygoid plates.

15. The ___ both have styloid processes. (a) scapula and clavicle; (b) malleus and incus; (c) tibia and fibula; (d) radius and temporal bones; (e) humerus and femur

16. The bone(s) that lack(s) a body or centrum is/are the: (a) first lumbar vertebra; (b) sacral vertebrae; (c) first cervical vertebra; (d) 12 thoracic vertebrae; (e) axis.

17. All of the following are paranasal sinuses *except* the: (a) maxillary sinus; (b) frontal sinus; (c) temporal sinus; (d) ethmoid sinus; (e) sphenoid sinus.

18. The lateral malleolus is a process of the: (a) humerus; (b) ulna; (c) radius; (d) fibula; (e) tibia.

19. The joint between the ___ is marked by an auricular surface on each bone. (a) atlas and axis; (b) last thoracic and first lumbar vertebrae; (c) clavicle and sternum; (d) hip bone and sacrum; (e) coccyx and sacrum

20. Of the eight carpal bones, the ___ is easily recognized by its distinctive hook. (a) pisiform; (b) hamate; (c) triquetral; (d) lunate; (e) navicular

21. Which of these is/are associated with the appendicular skeleton?
    1. intercondylar fossa
    2. calcaneus
    3. phalanges
    4. transverse process

    (a) 1 & 3; (b) 2 & 4; (c) 1, 2, & 3; (d) 4 only; (e) all the above

22. Which of these is/are correct articulation(s)?
    1. coracoid process, mandible/mandibular fossa, temporal bone
    2. manubrium/clavicle
    3. acromion, scapula/humeral head
    4. tubercle of rib/transverse costal facets, thoracic vertebrae

    (a) 1 & 3; (b) 2 & 4; (c) 1, 2, & 3; (d) 4 only; (e) all the above

23. Which of these is/are true about the interactions between the skeletal system and other organ systems?
    1. Bones provide calcium for normal muscle and neural function.
    2. Hormones regulate blood calcium levels and bone growth.
    3. Some bones of the skull form respiratory passages.
    4. The kidneys help to regulate blood calcium and phosphate levels.

    (a) 1 & 3; (b) 2 & 4; (c) 1, 2, & 3; (d) 4 only; (e) all the above

24. Which of these is/are true?
    1. The hard palate allows mammals to chew food and breathe at the same time.
    2. Anosmia is caused by a break in the sella turcica.
    3. Kyphosis is often a result of osteoporosis.
    4. Bulging fontanels of a newborn indicate severe hydrocephalus.

    (a) 1 & 3; (b) 2 & 4; (c) 1, 2, & 3; (d) 4 only; (e) all the above

25. Skeletal adaptations for bipedal locomotion include:
    1. the distinct shape of the arch of the human foot.
    2. the inferior placement of the foramen magnum of the occipital bone.
    3. the shape and angle of the hip bones.
    4. the nearly vertical position of the femur in the human thigh.

    (a) 1 & 3; (b) 2 & 4; (c) 1, 2, & 3; (d) 4 only; (e) all the above

E. Word Origins

1. In *dura mater, dura* means "deepest."
2. In *coronal, corona-* means "crown."
3. In *temporal, tempor-* means "side."
4. In *mastoid, mast-* means "lump."
5. In *acromion, acr-* means "peak, extremity."
6. In *scapula, scap-* means "scrape."
7. In *pterygoid, pteryg-* means "wing."
8. In *crista galli, crista* means "groove."
9. In *zygomatic, zygo-* means "joined."
10. In *costal, costa-* means "shoreline, edge."
11. lamina-
12. clav-
13. supra-
14. ante-
15. capit-

F. Which One Does Not Belong?

1. (a) radial notch; (b) trochlear notch; (c) styloid process; (d) intertubercular sulcus

2. (a) stylomastoid foramen; (b) petrous part; (c) foramen magnum; (d) carotid canal

3. (a) greater trochanter; (b) radial tuberosity; (c) deltoid tuberosity; (d) trochlear notch

4. (a) intercondylar notch; (b) intertrochanteric crest; (c) linea aspera; (d) trochlea

5. (a) sternum; (b) clavicle; (c) ulna; (d) femur

G. Matching Articulations and Bone Markings

I. Match the bone region (A–V) that articulates with the bone and/or region listed in 1–15.

A. femoral head
B. capitulum, humerus
C. distal tibia
D. proximal tibia
E. trochlea, humerus
F. lateral edge, radial head
G. mandibular fossa
H. perpendicular plate, ethmoid bone
I. navicular
J. glenoid cavity, scapula
K. manubrium
L. mandible
M. occipital condyles
N. axis
O. gladiolus
P. transverse process, thoracic vertebrae
Q. sphenoid bone
R. hip bone
S. crista galli, ethmoid
T. xiphoid process
U. metatarsals
V. costal facets, bodies of vertebrae

1. Acetabulum
2. Atlas, superior articulating facet
3. Clavicle, sternal end
4. Femoral condyles
5. Lateral sacrum
6. Humerus, head
7. Mandibular condyle
8. Inferior manubrium

9. Radius, proximal head
10. Radial notch, ulna
11. Rib, tubercle
12. Talus (distal)
13. Talus (superior)
14. Trochlear notch, ulna
15. Vomer (superior)

II. Match the bone marking with the correct bone.

A. humerus B. femur C. hip bone D. radius E. scapula F. ulna

1. Acetabulum
2. Acromion
3. Anterior superior iliac spine
4. Axillary border
5. Capitulum
6. Coracoid process
7. Coronoid process
8. Glenoid cavity
9. Gluteal lines
10. Greater trochanter
11. Greater tubercle
12. Intercondylar notch
13. Intertrochanteric crest
14. Intertubercular sulcus
15. Inferior pubic ramus
16. Infraspinous fossa
17. Linea aspera
18. Patellar surface
19. Obturator foramen
20. Olecranon
21. Olecranon fossa
22. Sciatic notches
23. Trochlea
24. Trochlear notch

9   Joints

A. Short Answer

1. In a(n) ___ joint, two bones are separated by a space filled with lubricating fluid.

2. Ulnar and radial flexion is hand movement in the ___ plane.

3. In a syndesmosis, two long bones are held together along most of their shafts by a sheet of fibrous tissue called the ___.

4. A collagenous cord or band that attaches a muscle to a bone is called a(n) ___, whereas one that attaches one bone to another is called a(n) ___.

5. In a(n) ___ joint, such as a metacarpophalangeal joint, an oval, convex surface on one bone inserts into a complementary depression on the other.

6. In a(n) ___ joint, such as the atlantoaxial joint, one bone has a round process held by a ligament against the other bone, and the first bone rotates with respect to the other.

7. When you make a fist, you are ___ the interphalangeal joints.

8. When you press on the gas pedal of a car or stand on tiptoes to reach something on a high shelf, your foot is making a motion called ___.

9. When you hold out your palm to receive something from another person, your forearm makes a rotational movement called ___.

10. The flexibility of a joint is measured in terms of its ___, the angle through which it can move.

11. The atlanto-occipital joint and movement of the talus on the tibia are examples of ___ levers.

12. The glenoid labrum is a cartilaginous ring that somewhat deepens the ___ joint.

13. Kinesiology is the study of ___.

14. The anterior and posterior cruciate ligaments are found deep within the cavity of the ___ joint.

15. Many joints have sacs filled with synovial fluid, called ___, that ease or redirect the actions of tendons.

B. Matching

A. meniscus	G. inversion	M. second-class	S. synovial membrane
B. lap	H. hyperextension	N. gomphosis	T. jaw
C. abduction	I. third-class	O. adduction	U. plane
D. ankle	J. flexion	P. synchondrosis	V. elbow
E. first-class	K. knee	Q. plane	W. synovial fluid
F. vitreous fluid	L. bursa	R. ball-and-socket	X. temporomandibular

1. Movement unique to the foot
2. Joint that involves condylar hinge, and plane elements
3. Lubricating fluid in a hinge joint
4. Cartilage pad between the articulating surfaces of two bones
5. Joint in which the round head of one bone fits into a cup-like depression of another
6. Movement of the neck when you stand and look up at the sky
7. Movement in which you go from a spread-legged position to anatomical position
8. Any joint with the fulcrum at one end, resistance at the other, and effort applied in the middle
9. Joint with anular ligaments
10. Joint with a lateral and medial meniscus

C. True or False

1. Diarthroses move more freely than amphiarthroses.

2. The tibial collateral ligament is medial in the knee joint.

3. A cartilaginous joint can be either a diarthrosis or a synarthrosis.

4. Teeth are not bones.

5. When you bite into an apple, your mandible is retracted.

6. If you trace a circle in the dirt with your toe, you are circumducting the leg.

7. The fusion of epiphysis and diaphysis in an adult long bone is an example of a synostosis.

8. All muscles whose tendons form the rotator cuff are on the posterior side of the scapula.

9. The articular cartilage of a bone is covered by a fibrous periosteum.

10. Hinge joints are biaxial, since they can be flexed and extended.

D. Multiple Choice

1. Touching your thumb to your fingertips is called: (a) supination; (b) protraction; (c) opposition; (d) extension; (e) abduction.

2. A bursa is most nearly related to a: (a) suture; (b) gomphosis; (c) ligament; (d) meniscus; (e) tendon sheath.

3. An intervertebral disc is an example of a: (a) symphysis; (b) meniscus; (c) synarthrosis; (d) fibrous joint; (e) condylar joint.

4. As you take a bite from an apple, you are using mainly ___ of the mandible. (a) protraction and retraction; (b) inversion and eversion; (c) flexion and extension; (d) adduction and abduction; (e) elevation and depression

5. The trapezium and metacarpal I form a unique ___ responsible for the unusual range of motion of the thumb. (a) synarthrosis; (b) gomphosis; (c) plane joint; (d) hinge joint; (e) saddle joint

6. You are driving your car, and you raise your right arm to rest it on the back of the seat. This motion is: (a) elevation; (b) abduction; (c) adduction; (d) pronation; (e) hyperextension.

7. You are driving your car, and you lift your foot slightly from the gas pedal to slow down. This motion is: (a) dorsiflexion; (b) plantar flexion; (c) inversion; (d) eversion; (e) elevation.

8. An immovable joint is a: (a) synarthrosis; (b) diarthrosis; (c) symphysis; (d) syndesmosis; (e) synovial joint.

9. Gout is a condition in which: (a) bones become displaced in a joint; (b) a joint prosthesis fails; (c) osteoarthritis occurs; (d) uric acid crystals accumulate in joints; (e) the immune system attacks the joint capsule.

10. The round ligament or ligamentum teres inserts on or into the: (a) glenoid cavity; (b) scapular notch; (c) fovea capitis; (d) semilunar notch; (e) pubic arch.

11. An example of a gomphosis is: (a) a tooth in its socket; (b) the coxal joint; (c) interosseous joints; (d) a lap suture; (e) intervertebral discs.

12. A ___ can never have a mechanical advantage greater than 1.0 because the resistance is always farther away from the fulcrum than the effort is. (a) hinge joint; (b) first-class lever; (c) second-class lever; (d) third-class lever; (e) pivot joint

13. In a(n) ___, two adjoining bones have interlocking, wavy edges. (a) amphiarthrosis; (b) fibrous joint; (c) plane suture; (d) serrate suture; (e) lap suture

14. A man is standing in line at a movie theater when he hyperextends his shoulder joint. Most likely, he did this in order to: (a) reach down and tie his shoe; (b) reach behind him to get his wallet from his hip pocket; (c) place his money on the counter; (d) clasp his hands together in front of him; (e) scratch the top of his head.

15. The glenoid labrum of the shoulder serves the same purpose as the ___ labrum of the hip. (a) humeral; (b) femoral; (c) acetabular; (d) sacroiliac; (e) pubofemoral

16. The pubic symphysis is a: (a) synovial joint; (b) cartilaginous joint; (c) gomphosis; (d) synarthrosis; (e) diarthrosis.

17. In woodworking, a miter joint is formed by two boards overlapping with beveled (sloping) edges. In the skull, the analogous type of joint is a: (a) lap suture; (b) plane suture; (c) coronal suture; (d) lambdoid suture; (e) serrate suture.

18. When you open your refrigerator door, bringing your hand toward your naval, you are ___ your arm/humerus. (a) everting; (b) protracting; (c) abducting; (d) internally rotating; (e) hyperextending

19. Opposition is unique to the: (a) hip joint; (b) wrist; (c) thumb; (d) elbow; (e) foot.

20. The subdeltoid, subacromial, and subcoracoid bursae are components of the ___ joint. (a) knee; (b) ankle; (c) first metacarpophalangeal; (d) atlanto-occipital; (e) shoulder

21. A meniscus can be found in which of the following joint(s)?
    1. glenohumeral
    2. elbow
    3. sternoclavicular
    4. knee

    (a) 1 & 3; (b) 2 & 4; (c) 1, 2, & 3; (d) 4 only; (e) all the above

22. Which of these is/are true about joints?
    1. The biceps brachii tendon is the most important stabilizer of the shoulder joint.
    2. The iliofemoral ligament is found in the hip joint.
    3. The elbow includes both pivot and hinge joints.
    4. The ankle joint includes the distal tibia and fibula articulating with the talus of tarsals.

    (a) 1 & 3; (b) 2 & 4; (c) 1, 2, & 3; (d) 4 only; (e) all the above

23. Which of these is/are true about joint problems?
    1. Osteoarthritis is caused by disuse of the joints.
    2. Gout is most common in women who eat a diet rich in protein.
    3. Bursitis is often caused by arthritis.
    4. Rheumatoid arthritis is an autoimmune disease found more often in women than in men.

    (a) 1 & 3; (b) 2 & 4; (c) 1, 2, & 3; (d) 4 only; (e) all the above

24. Which of these is/are true?
    1. In arthroplasty, artificial joints replace diseased joints.
    2. Joint dislocation is most often caused by fractures.
    3. Arthroscopy has improved diagnosis and treatment of joint injuries.
    4. Locking the knee in order to stand for long periods involves mainly the medial and lateral collateral ligaments.

    (a) 1 & 3; (b) 2 & 4; (c) 1, 2, & 3; (d) 4 only; (e) all the above

25. Which of these affect(s) ROM?
    1. shape of bone surfaces in a joint
    2. muscle tendons that cross a joint
    3. tone of the muscles whose tendons are involved in a joint
    4. strength of ligaments at a joint

    (a) 1 & 3; (b) 2 & 4; (c) 1, 2, & 3; (d) 4 only; (e) all the above

E. Word Origins

1. In *diarthrosis, arthr-* means "disconnected."
2. In *synostosis, syn-* means "together."
3. In *symphysis, sym-* means "simple."
4. In *synovial, ovi-* means "egg."
5. In *abduction, ab-* means "away."
6. In *supinate, supin-* means "to move around."
7. In *glenoid labrum, labrum* means "lip."
8. In *ligamentum teres, teres* means "delicate."
9. In *cruciate, cruci-* means "nail."
10. In *adduction, duc-* means "lead."
11. burs-
12. amphi-
13. gompho
14. -physis
15. -iscus

F. Which One Does Not Belong?

1. (a) articular cartilage; (b) fibrous capsule; (c) synovial fluid; (d) synostosis

2. (a) radioulnar; (b) intervertebral; (c) humeroulnar; (d) intercarpal

3. (a) gomphosis; (b) suture; (c) hinge; (d) syndesmosis

4. (a) tibial collateral; (b) anterior cruciate; (c) transverse; (d) teres

5. (a) circumduction; (b) plantar flexion; (c) eversion; (d) dorsiflexion

G. Figure Exercise

Match the structures in the figures with the tendons or ligaments.

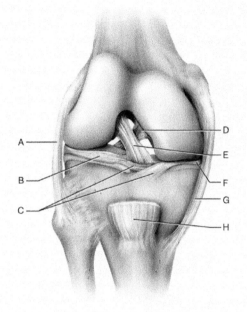
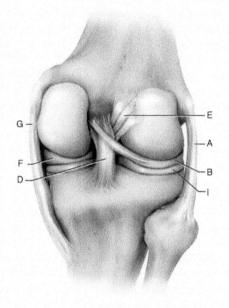

1. Anterior cruciate ligament
2. Tibial collateral ligament
3. Medial meniscus
4. Fibular collateral ligament
5. Lateral meniscus
6. Posterior cruciate ligament
7. Transverse ligament

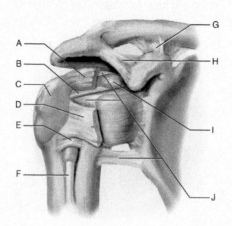

8. Biceps brachii tendon
9. Coracoacromial ligament
10. Coracoclavicular ligament
11. Coracohumeral ligament
12. Glenohumeral ligament
13. Subscapularis tendon
14. Supraspinatus tendon
15. Transverse humeral ligament

10   The Muscular System

A.   Short Answer

1. A(n) _Sphincter_ is a circular muscle that controls a body opening.

2. The point of convergence and divergence of several facial muscles is called the _Modiolus_.

3. In a(n) _Fusiform_ muscle, the fascicles converge toward each end, so the ends of the muscle are tapered and there is a thick belly.

4. The _Prime mover_ at a given joint is the muscle that is mainly responsible for a given motion, while a(n) _antagonist_ is a muscle that opposes that action.

5. A(n) _Intrinsic_ muscle is one found entirely within a certain region of study, such as the hand or eye, while a(n) _Extrinsic_ muscle is one that acts on that region but originates somewhere outside of it.

6. The _zygomaticus_ major and minor draw the corners of the mouth upward when you smile.

7. The ___ and ___ muscles elevate the mandible when you take a bite of food. _Temporalis & masseter_

8. In elbow flexion, the rhomboid muscles act as _fixators_ to prevent scapular movement.

9. The muscles most involved in inhaling are the _external intercostal_ between the ribs and the _diaphram_ between the thoracic and abdominal cavities.

10. The medial, superficial muscle of the abdomen, extending vertically from pubis to sternum, is the ___. _Rectus Abdominal_

11. The large, four-headed muscle of the anterior aspect of the thigh is the _Quadricet femoris_

12. The triangular muscle of the shoulder, where injections are often given, is the _Deltiod_

13. Tendons of the infraspinatus, supraspinatus, subscapularis, and teres minor muscles form the _rotator cuffs_, a common site of sports injuries.

14. The synergist of the biceps brachii, just deep to it, is the _brachialis_

15. Muscles of the head and neck are innervated by _Cranial_ nerves, whereas most muscles of the body are innervated by _Spinal_ nerves.

B. Matching

A. hamstrings	G. deltoid	M. masseter	S. pennate
B. pterygoids	H. sartorius	N. soleus	T. biceps brachii
C. temporalis	I. prime mover	O. scalenes	U. anterior compartment
D. sartorius	J. splenius capitis	P. digastric	V. quadriceps femoris
E. triangular	K. fusiform	Q. synergist	W. origin
F. adductor pollicis	L. brachialis	R. insertion	X. aponeurosis

1. Longest muscle in the body
2. The movable end of a muscle
3. Feather-like muscle with fascicles converging on a central tendon
4. Muscle that aids or modifies the action of a prime mover
5. Produce(s) lateral excursions of the mandible during chewing
6. Abduct(s) the arm
7. Rotate(s) the head
8. Synergist of the gastrocnemius
9. A muscle of the thenar eminence
10. Includes the vastus lateralis, medialis, and intermedius

C. True or False

1. The same muscle can serve as a prime mover in one joint action and an antagonist in another.

2. The most superficial muscle of the anterior cervical region is the platysma.

3. The floor of the mouth is formed mainly by the omohyoid muscle.

4. The large, superficial muscle of the upper back is the latissimus dorsi.

5. The transverse abdominal is sandwiched between the internal oblique and external oblique.

6. As we age, our muscles lose tone and mass which can lead to an increased risk of type 2 diabetes mellitus.

7. The subscapularis covers most of the posterior surface of the scapula inferior to its spine.

8. The triceps brachii is a synergist of the biceps brachii.

9. One end of a muscle can function as its origin during one action and can function as its insertion during a different action.

10. The extensor hallucis longus and flexor hallucis longus are extrinsic muscles of the foot.

D. Multiple Choice

1. If you see the word ___ in the name of a muscle, you can deduce that it must have something to do with the great toe. (a) *pollicis;* (b) *hallucis;* (c) *lumbrical;* (d) *fibularis;* (e) *indicis*

2. If you see the word *biceps* in the name of a muscle, you can deduce that it must have: (a) something to do with the arm; (b) something to do with the head; (c) two heads; (d) two insertions; (e) two fascicles.

3. Which of the following muscles is *not* in the suprahyoid group? (a) geniohyoid; (b) sternohyoid; (c) digastric; (d) mylohyoid; (e) stylohyoid

4. Which of the following muscles does *not* produce expressions of the mouth? (a) risorius; (b) depressor labii inferioris; (c) zygomaticus; (d) corrugator supercilii; (e) mentalis

5. The medial and lateral pterygoid muscles insert on the: (a) hyoid bone; (b) occipital bone; (c) sternum and clavicle; (d) maxilla; (e) mandible.

6. In humans, the most useless of the following muscles is/are the: (a) masseter; (b) auricularis; (c) superior and inferior obliques; (d) genioglossus; (e) trapezius.

7. The connective tissue sheath that encloses an entire muscle is called: (a) the endomysium; (b) a fascicle; (c) the fascia; (d) the epimysium; (e) the sarcolemma.

8. When you hyperextend your wrist, as if admiring a new ring, you can perhaps see four prominent tendons on the back of your hand leading to digits II–V. These tendons belong to the: (a) flexor digitorum superficialis; (b) flexor digitorum profundus; (c) lumbrical muscles II–V; (d) extensor digitorum; (e) dorsal interosseous muscles II–V.

9. Which of the following is *not* a muscle of the lower extremity? (a) sartorius; (b) adductor magnus; (c) pectineus; (d) semitendinosus; (e) flexor carpi ulnaris

10. The prominent tendons you can palpate on each side of the popliteal fossa arise from the: (a) hamstring muscles; (b) triceps brachii; (c) biceps brachii; (d) quadriceps femoris; (e) gastrocnemius.

11. Crossing your legs while you are sitting employs mainly the: (a) sartorius; (b) superior gemellus; (c) piriformis; (d) pectineus; (e) ischiocavernosus.

12. Which of the following muscles would not be externally visible on the torso of an Olympic swimmer? (a) latissimus dorsi; (b) teres major; (c) transverse abdominal; (d) trapezius; (e) serratus anterior

13. The calcaneal (Achilles) tendon arises from the: (a) gastrocnemius and soleus; (b) fibularis longus and fibularis brevis; (c) sartorius and biceps femoris; (d) semitendinosus and semimembranosus; (e) adductor longus, adductor brevis, and adductor magnus.

14. Which of these structures is *not* a muscle? (a) extensor digitorum; (b) extensor retinaculum; (c) extensor carpi ulnaris; (d) extensor hallucis longus; (e) extensor digiti minimi

15. The teres major and teres minor: (a) originate on the humerus and insert on the scapula; (b) originate on the scapula and insert on the humerus; (c) originate on the spinal column and insert on the scapula; (d) originate on the hip bone and insert on the femur; (e) originate on the tibia and insert on the calcaneus and talus.

16. The word ___ in the name of a muscle suggests that it is associated with the fingers or toes. (a) *pronator;* (b) *quadriceps;* (c) *intercostal;* (d) *digitorum;* (e) *rhomboideus*

17. Tennis elbow is an inflammation of the: (a) biceps brachii insertion; (b) origin of the flexor carpi ulnaris; (c) extensor pollicis longus; (d) brachioradialis insertion; (e) origin of the extensor carpi muscles.

18. When the head is slightly turned and the neck is tensed, a thick, ropy muscle called the ___ can be seen extending from the sternum to a point just behind the ear. (a) trapezius; (b) latissimus dorsi; (c) sternocleidomastoid; (d) anterior scalene; (e) splenius capitis

19. The ___ muscles converge from the ischial tuberosities to the penis or clitoris and aid in erection. (a) ischiocavernosus; (b) bulbospongiosus; (c) transverse perineal; (d) coccygeus; (e) pubococcygeus

20. A(n) ___ is a broad, sheet-like tendon. (a) ligament; (b) aponeurosis; (c) interosseous membrane; (d) fascia; (e) epimysium

21. Which of these is/are true?
    1. The galea aponeurotica lies superior to both the frontal and occipital bones.
    2. The modiolus is found on the elbow.
    3. The erector spinae group includes the longissimus.
    4. A muscle that pronates the hand moves it into anatomical position.

    (a) 1 & 3; (b) 2 & 4; (c) 1, 2, & 3; (d) 4 only; (e) all the above

22. Which of these is/are used in producing facial expressions?
    1. platysma
    2. levator anguli oris
    3. occipitofrontalis
    4. masseter

    (a) 1 & 3; (b) 2 & 4; (c) 1, 2, & 3; (d) 4 only; (e) all the above

23. Which of these has/have an attachment(s) to the femur?
    1. external oblique
    2. gastrocnemius
    3. fibularis brevis
    4. piriformis

    (a) 1 & 3; (b) 2 & 4; (c) 1, 2, & 3; (d) 4 only; (e) all the above

24. Muscles of the medial compartment of the thigh include the:
    1. pectineus.
    2. adductor magnus.
    3. gracilis.
    4. sartorius.

    (a) 1 & 3; (b) 2 & 4; (c) 1, 2, & 3; (d) 4 only; (e) all the above

25. Which of these is/are true?
    1. Inguinal hernias are often seen in swimmers.
    2. Carpal tunnel syndrome is an occupational hazard for people who play racquet sports.
    3. When one suffers a pulled groin, the muscle most affected is the vastus lateralis.
    4. Intramuscular (IM) injections are most common in the deltoid and in the gluteus medius.

    (a) 1 & 3; (b) 2 & 4; (c) 1, 2, & 3; (d) 4 only; (e) all the above

E. Word Origins

1. In *perimysium, mys-* means "muscle."
2. In *splenius capitis, capit-* means "head."
3. In *opponens pollicis, pollicis* means "of the thumb."
4. In *levator labii, labi-* means "tongue."
5. In *orbicularis oris, oris* means "round."
6. In *genioglossus, genio-* means "chin."
7. In *digastric, gastr-* means "belly."
8. In *semispinalis cervicis, cervic-* means "circular."
9. In *biceps, -ceps* means "head."
10. In *extensor hallucis, hallucis* means "of the big toe."
11. cleido-
12. rect-
13. sartor-
14. bucc-
15. -phragm

F.   Which One Does Not Belong?

1. (a) adductor pollicis brevis; (b) abductor digiti minimi; (c) flexor pollicis longus; (d) lumbricals

2. (a) semitendinosus; (b) biceps femoris; (c) semimembranosus; (d) gracilis

3. (a) biceps; (b) carpi; (c) cervicis; (d) abdominis

4. (a) heat production; (b) flexion; (c) movement; (d) communication

5. (a) temporalis; (b) medial pterygoid; (c) digastric; (d) masseter

G. Figure Exercises

I. First label the diagram on the following page; then answer the following questions.

1. Which of these are the hamstring muscles?
2. Which of these muscles inserts on the clavicle, acromion, and scapular spine?
3. Identify the rotator cuff muscles. (Only three of the four are shown here.)
4. This muscle is the antagonist to the biceps brachii.
5. This muscle is the "boxer's muscle" originating on ribs 1–9, inserting on the medial border of the scapula.
6. This muscle flexes the knee and plantar flexes the foot.
7. Which two muscles flex the knee, extend the hip, and medially rotate the tibia?
8. Which muscles originate on the humerus? (Only four are shown in this diagram.)
9. Which muscle provides most of the power when climbing stairs?
10. This muscle is called the "swimmer's muscle" because of its action on the humerus.

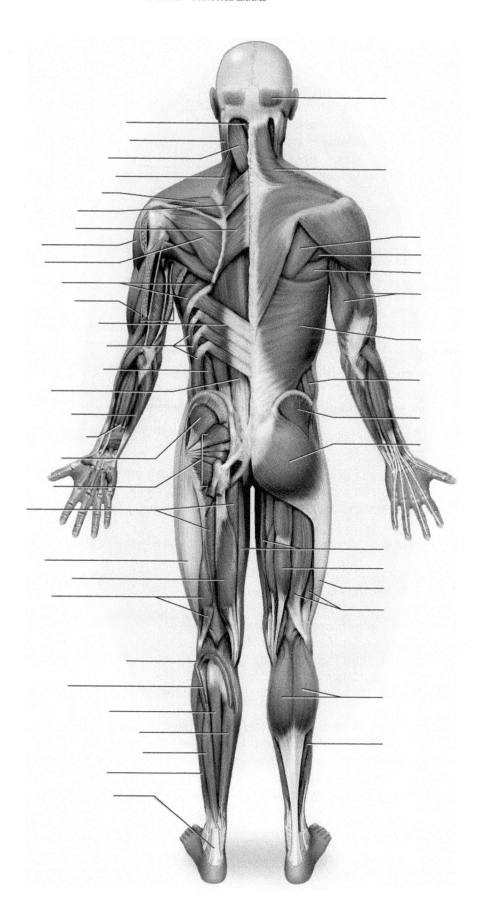

II. Match the muscles with their attachments on these bones. Some muscles may have more than
     one answer.

A. radial tuberosity
B. acromion, scapula
C. fibular head
D. medial border, scapula
E. styloid process, radius
F. olecranon process, ulna

G. anterior superior iliac spine
H. posterolateral femur
I. lateral condyle, femur
J. calcaneus
K. intertubercular sulcus, humerus
L. lesser trochanter, femur

M. greater trochanter, femur
N. pubis
O. ischial tuberosity, hip bone
P. iliac fossa
Q. lesser tubercle, humerus
R. posterior, inferior angle, scapula

1. Biceps brachii
2. Brachioradialis
3. Deltoid
4. Gastrocnemius
5. Gluteus medius
6. Latissimus dorsi
7. Semitendinosus
8. Iliopsoas
9. Rectus abdominis
10. Sartorius
11. Serratus anterior
12. Subscapularis
13. Teres major
14. Triceps brachii
15. Vastus lateralis

III. Name at least 10 muscles that have attachments on the scapula.

11 Muscle Cells

A. Short Answer

1. Action potentials are propagated to the interior of a muscle fiber by extensions of the sarcolemma called the ___.

2. The ___ serves as a reservoir of calcium ions in skeletal muscle.

3. The thick myofilaments are composed of the protein ___.

4. Cardiac and some smooth muscle tissue do not require direct stimulation by the nervous system; thus, they are said to be ___.

5. The synapse where a motor nerve fiber and a skeletal muscle fiber meet is called a(n) ___.

6. Motor nerve fibers stimulate skeletal muscle fibers with a neurotransmitter called ___.

7. At rest a skeletal muscle fiber has a potential of about -90 mV, called its ___, across the sarcolemma.

8. When a muscle fiber is excited, its membrane produces a brief, self-propagating voltage change called a(n) ___.

9. Currently, our best understanding of how muscle contracts is called the ___ theory.

10. Myosin cannot bind to actin until calcium binds to ___ and ___ moves out of the way of the active sites on actin.

11. When a muscle is stimulated so frequently it can't completely relax between twitches, the successive twitches become stronger and stronger. This is called ___.

12. In ___ contraction, a muscle develops or maintains tension even as it is being stretched, thus preventing it from relaxing too quickly and producing abrupt, uncoordinated actions.

13. The difference between one's normal rate of oxygen consumption and the rate seen at the end of a strenuous exercise is called ___.

14. In ___ smooth muscle, neighboring muscle cells are joined by gap junctions and stimulate each other to contract.

15. Most sympathetic nerve fibers stimulate smooth muscle contraction via the neurotransmitter ___.

B. Matching

A. dystrophin	G. isotonic contraction	M. glycogen	S. tetanus
B. calmodulin	H. carbon dioxide	N. caveolae	T. myoglobin
C. T-tubules	I. Z discs	O. myosin	U. calsequestrin
D. twitch	J. tropomyosin	P. treppe	V. recovery stroke
E. synaptic cleft	K. isometric contraction	Q. A band	W. eccentric contraction
F. creatine	L. phosphagen system	R. lactic acid	X. synaptic vesicles

1. Structures that define the limits of a sarcomere
2. End product of anaerobic fermentation
3. Large protein linking actin in thin myofilaments to sarcolemma proteins
4. Calcium-binding protein of the sarcoplasmic reticulum
5. Source of stored energy for muscle contraction
6. Location of acetylcholinesterase
7. Pockets in smooth muscle myocytes that increase the surface area and number of calcium channels
8. Sustained muscle contraction at a high-stimulus frequency
9. Shortening of a muscle while maintaining constant tension
10. Provides energy for short bursts of skeletal muscle activity

C. True or False

1. When a muscle contracts, each thin filament gets shorter.

2. Skeletal muscle cannot contract unless it is stimulated by a motor neuron or is artificially stimulated.

3. Tropomyosin is part of the thin filaments of the sarcomeres.

4. A "large" motor unit is one with many nerve fibers per muscle fiber.

5. Myofilaments contain myofibrils laid end to end in a sarcomere.

6. When ATP is unavailable, creatine phosphate can bind to myosin and serve in place of ATP.

7. A muscle cell's resting membrane potential is maintained by the Na^+–K^+ pump.

8. It would be impossible to record a resting potential in a cell if both electrodes were inserted into the cytoplasm.

9. Muscles engage in anaerobic respiration when exercise lasts more than about 10 minutes.

10. The A bands of skeletal muscle contain only myosin.

D. Multiple Choice

1. Creatine kinase: (a) donates one of its P_i groups to ADP; (b) phosphorylates and activates certain cytoplasmic enzymes; (c) acts as a second messenger in muscle cells; (d) catalyzes the transfer of P_i from CP to ADP; (e) functions as a substitute for ATP during anaerobic respiration.

2. Which of the following is *not* found in the thin filaments of skeletal muscle? (a) F actin; (b) ATPase; (c) troponin; (d) tropomyosin; (e) Ca^{2+} receptors

3. Muscle contraction and relaxation require ATP for all of the following processes *except:* (a) flexion of the head of a myosin molecule; (b) maintenance of the excitability of the sarcolemma; (c) release of Ca^{2+} from terminal cisternae; (d) reabsorption of Ca^{2+} from the sarcoplasm; (e) use of Na^+–K^+ pumps to restore resting membrane potential.

4. In contrast to slow oxidative muscle fibers, fast glycolytic fibers: (a) contract more slowly; (b) contain more mitochondria; (c) fatigue more quickly; (d) have more capillaries; (e) contain more myoglobin.

5. When a muscle develops tension but does not shorten, it is said to exhibit: (a) treppe; (b) fatigue; (c) twitch; (d) isometric contraction; (e) isotonic contraction.

6. Endurance training has the *least* effect on the: (a) thickness of muscle fibers; (b) number of mitochondria in a muscle fiber; (c) amount of glycogen in a muscle fiber; (d) red blood cell count; (e) density of blood capillaries in a muscle.

7. Autonomic nerve fibers release neurotransmitters from varicosities in: (a) single-unit smooth muscle; (b) multi-unit smooth muscle; (c) neuromuscular junctions; (d) skeletal muscles; (e) type II (fast-twitch) fibers.

8. In order for a skeletal muscle to relax, ___ must be enzymatically degraded. (a) lactic acid; (b) myosin; (c) acetylcholine; (d) acetylcholinesterase; (e) calcium

9. In skeletal and cardiac muscle, actin and myosin overlap in the ___ of a sarcomere. (a) Z disc; (b) H band; (c) dark band; (d) light band; (e) triads

10. A skeletal muscle fiber has more ___ than it does any of these other features. (a) myofibers; (b) sarcomeres; (c) neuromuscular junctions; (d) synaptic vesicles; (e) H zones

11. When there is not enough oxygen to produce ATP by aerobic respiration, a muscle fiber can produce some ATP by borrowing phosphate groups from: (a) adenosine triphosphate; (b) creatine phosphate; (c) creatine kinase; (d) myoglobin; (e) ACh.

12. Which of these is present in smooth muscle myocytes but not in myofibers of skeletal muscle? (a) calmodulin; (b) actin; (c) mitochondria; (d) myosin ATPase; (e) endoplasmic reticulum

13. A ___ contains sacomeres, glycogen, and myoglobin. (a) myosin; (b) triad; (c) myofiber; (d) myofilament; (e) myofibril

14. Which of the following events occurs *first* at a neuromuscular junction? (a) Myosin heads bind to receptor sites of G actin. (b) Acetylcholine is released into the synaptic cleft. (c) An action potential travels down a motor neuron. (d) Ca^{2+} is released from the sarcoplasmic reticulum. (e) Na^+–K^+ gates open in the sarcolemma.

15. The term *motor unit* refers to: (a) a neuromuscular junction; (b) the distance from one Z disc to the next; (c) one thick filament and all the thin filaments with which it forms cross-bridges; (d) one nerve fiber and all the muscle fibers it innervates; (e) one myofibril of a muscle fiber.

16. Actin is lacking from the ___ of a relaxed sarcomere. (a) H band; (b) I band; (c) A band; (d) thin filaments; (e) myofibrils

17. Based on the sliding filament theory, we would expect ___ during skeletal muscle contraction. (a) the A and H bands to disappear; (b) the I bands to get shorter and A bands to remain the same length; (c) the A bands to get shorter and I bands to remain the same length; (d) the A and I bands both to get shorter; (e) the H bands and Z discs to disappear

18. The -90 mV resting membrane potential of a skeletal muscle cell is maintained by the: (a) fact that K^+ is much more concentrated in the ECF than in the ICF; (b) leakage of Na^+ into the cell; (c) action of the Na^+–K^+ pump; (d) presence of large, nondiffusible anions in the ICF; (e) action of ACh on the sarcolemma.

19. Consider the following events that are involved in muscle contraction and relaxation, although they do not necessarily occur in this order, and some steps are omitted:
    I. Ca^{2+} enters the synaptic knob.
    II. Ca^{2+} is released by the sarcoplasmic reticulum.
    III. Ca^{2+} is pumped into the sarcoplasmic reticulum.
    IV. Ca^{2+} binds to troponin.
    V. Tropomyosin moves away from the myosin binding sites on F actin.
    Which of the following correctly represents the order in which these events occur?
    (a) I-II-III-IV-V; (b) V-IV-III-II-I; (c) II-I-IV-V-III; (d) III-II-I-IV-V; (e) I-II-IV-V-III

20. In which of these would you expect there to be a greater proportion of fast glycolytic muscle fibers? (a) swimmers; (b) marathon runners; (c) cyclists; (d) cross-country skiers; (e) basketball players

21. Which of these is/are true about cardiac and smooth muscle?
    1. They both have gap junctions through which action potentials are carried.
    2. They have pacemakers.
    3. They use Ca^{2+} to initiate the contraction process.
    4. They are innervated by somatic motor nerve fibers.

    (a) 1 & 3; (b) 2 & 4; (c) 1, 2, & 3; (d) 4 only; (e) all the above

22. Which of these is/are true about muscle disorders?
    1. Myesthenia gravis can be treated with acetylcholinesterase inhibitors.
    2. Crush syndrome is often confused with fibromyalgia.
    3. In muscular dystrophy, muscles become weak and are replaced by fibrous and adipose tissues.
    4. Most muscular dystrophies are caused by autosomal-dominant genes.

    (a) 1 & 3; (b) 2 & 4; (c) 1, 2, & 3; (d) 4 only; (e) all the above

23. Malathion is an organophosphate pesticide that blocks the action of acetylcholinesterase. Which of these would be expected as a result of exposure to this pesticide?
    1. intermittent tetanus
    2. flaccid paralysis
    3. increased number of slow oxidative fiber stimulation
    4. continual contraction of muscles or cholinergic crisis

    (a) 1 & 3; (b) 2 & 4; (c) 1, 2, & 3; (d) 4 only; (e) all the above

24. Which of these is/are true about muscle structure?
    1. Myofibrils in skeletal muscle are composed of numerous fascicles.
    2. Both troponin and calmodulin bind calcium ions.
    3. Thick filaments contain T-tubules that are associated with the sarcoplasmic reticulum.
    4. Sarcomeres are present in skeletal and cardiac muscle, but not in smooth muscle tissue.

    (a) 1 & 3; (b) 2 & 4; (c) 1, 2, & 3; (d) 4 only; (e) all the above

25. Which of these is/are true about the causes of muscle fatigue?
    1. Lactic acid accumulation may interfere with calcium activity.
    2. Electrolyte loss through sweating can decrease sarcolemma excitability.
    3. An increase in ECF potassium hyperpolarizes muscle fiber membranes, making them less able to conduct action potentials.
    4. ATP depletion is the ultimate reason for fatigue to occur.

    (a) 1 & 3; (b) 2 & 4; (c) 1, 2, & 3; (d) 4 only; (e) all the above

E. Word Origins

1. In *sarcoplasmic, sarco-* means "full."
2. In *H-band,* helle means "bright."
3. In *isometric, iso-* means "same."
4. In *isometric, metr-* means "measuring device."
5. In *isotonic, ton-* means "heavy."
6. In *dystrophy, dys-* means "degenerate."
7. In *myoglobin, myo-* means "muscle."
8. In *acetylcholinesterase, -ase* means "enzyme."
9. *Treppe* means "time."
10. In *phosphagen, -gen* means "produce."
11. -lemma
12. mortis
13. -trophy
14. auto-
15. -blast

F. Which One Does Not Belong?

1. (a) junctional folds; (b) acetylcholine; (c) sarcoplasmic reticulum; (d) synaptic knob

2. (a) conductivity; (b) excitability; (c) elasticity; (d) autorhythmicity

3. (a) tropomyosin; (b) troponin; (c) calmodulin; (d) calsequestrin

4. (a) myosin light-chain kinase; (b) phosphagen system; (c) aerobic respiration; (d) anaerobic respiration

5. (a) striation; (b) calmodulin; (c) branched cells; (d) intercalated discs

G. Figure Exercise

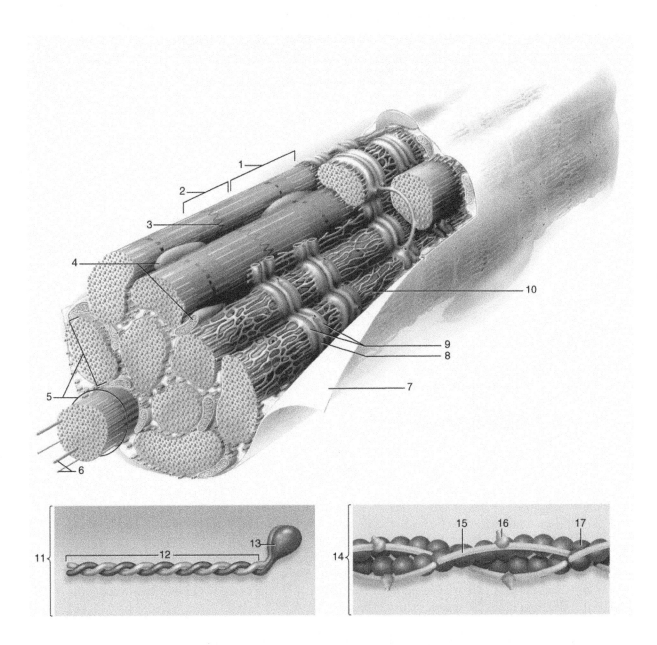

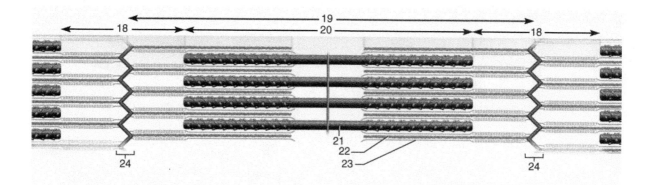

Using the figures above and on p. 99, answer the following questions about skeletal muscle cell structure and function.

1. Which of these are capable of carrying an electrical current or action potential?
   (a) 3 & 5; (b) 5 & 7; (c) 7 & 8; (d) 8 & 10; (e) 9 & 10

2. What structure stores calcium ions in resting muscle?
   (a) 6; (b) 7; (c) 8; (d) 9; (e) 10

3. Sarcomeres laid end to end make up these.
   (a) 2; (b) 5; (c) 6; (d) 8; (e) 10

4. Between two of these is a sarcomere.
   (a) 1; (b) 2; (c) 3; (d) 6; (e) 8

5. Which of these binds calcium ions before a muscle cell can contract?
   (a) 7; (b) 8; (c) 17; (d) 13; (e) 16

6. Which of these binds to active sites on actin during muscle contraction?
   (a) 11; (b) 17; (c) 13; (d) 15; (e) 12

7. The area marked 19 is called: (a) a sarcomere; (b) a thin filament; (c) the light band;
   (d) a neuromuscular junction; (e) the dark band.

8. When a muscle contracts, which of these occurs?
   (a) 21 slides past 22. (b) 23 relaxes. (c) The length of 19 increases. (d) 13 attaches to and pulls on 17.
   (e) Calcium binds 22 and shortens 19.

9. The entire structure shown in 1–10 is a portion of a: (a) myofibril; (b) muscle filament;
   (c) myofiber; (d) myoblast; (e) thick filament.

10. Which of these is/are the dark band(s)?
    (a) 11 & 14; (b) 24; (c) 19; (d) 18; (e) 20

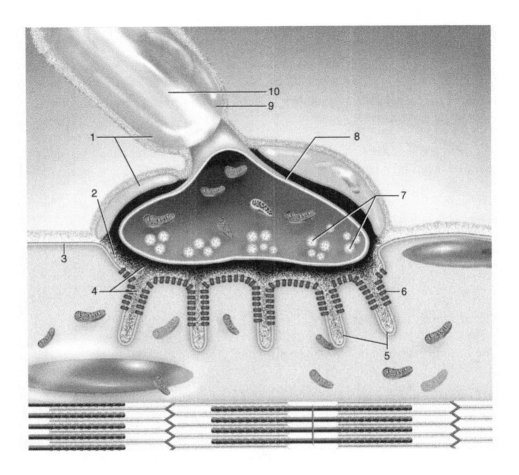

Answer the following questions about this synapse.

11. Which of these is/are capable of an action potential?
    (a) 5, 6, & 7; (b) 2, 5, & 8; (c) 1 & 10; (d) 8 & 9; (e) 3, 5, 8, & 10

12. ___ is the process by which #7 deposit their neurotransmitters to #2.
    (a) Diffusion; (b) Exocytosis; (c) Action potential; (d) Osmosis; (e) End plate potential

13. Which of these is capable of initiating end plate potentials on the sarcolemma?
    (a) 2; (b) 5; (c) 7; (d) 6; (e) 8

14. Which of these is continuous with T-tubules?
    (a) 3; (b) 8; (c) 2; (d) 4; (e) 6

15. Which of these contains voltage-gated calcium channels?
    (a) 1; (b) 2; (c) 8; (d) 6; (e) 10

12 Nerve Cells

A. Short Answer

1. Neurons are greatly outnumbered by ___ cells, which perform various supportive and protective roles in the nervous system.

2. The neurosoma of a neuron receives information from processes called ___ and transmits signals to the next cell by way of a process called the ___.

3. The axon hillock and initial segment of a neuron constitute its ___, so named because this is where action potentials are first generated.

4. Neurons that carry signals from the CNS to muscles and glands are called ___ neurons.

5. The ___ continues as long as sodium gates are open in a neuron membrane.

6. The myelin sheath is formed by ___ in the CNS and by ___ in the PNS.

7. The periodic gaps in a myelin sheath are called ___.

8. Neurotransmitters are released by the exocytosis of organelles called ___.

9. ___ is any shift in the membrane potential of a neuron toward 0 mV.

10. A small positive voltage change caused by a neurotransmitter is called a(n) ___.

11. For a neuron to fire, several of the voltage changes in question 10 must occur and add up to a minimum voltage called the ___.

12. The ability to process, store, and recall information using neurons is called ___, and it is based on postsynaptic potentials produced by neurotransmitters at synapses.

13. ___provide electrical insulation for neurosomas in ganglia of the PNS.

14. Name any two catecholamines.

15. ___ is a process in which one neuron can suppress the release of a neurotransmitter by another neuron.

B. Matching

A. neuron	G. IPSP	M. trigger zone	S. diverging circuit
B. gray matter	H. threshold	N. substance P	T. K^+ efflux
C. 0 mV	I. neuron doctrine	O. GABA	U. ganglia
D. converging circuit	J. glutamic acid	P. myelin sheath	V. terminal arborization
E. internodes	K. white matter	Q. astrocytes	W. all-or-none law
F. +30 mV	L. nerve	R. Na^+ influx	X. nodes of Ranvier

1. Location of all neurosomas in the CNS
2. Form scar tissue in damaged brain tissue
3. Site of most voltage-gated ion gates in a myelinated nerve fiber
4. Bundles of axons wrapped in a connective tissue sheath
5. Site of temporal summation
6. Causes repolarization in the second half of an action potential
7. Excitatory neurotransmitter in some brain regions
8. Clusters of neurosomas in the PNS
9. A neuropeptide
10. One presynaptic neuron producing output in multiple postsynaptic neurons

C. True or False

1. Each node of Ranvier speeds up the transmission of a nerve signal.

2. All neurons have their neurosomas located in the CNS.

3. The visceral motor division of the PNS includes the parasympathetic and sympathetic divisions.

4. People with parkinsonism cannot produce normal amounts of acetylcholine.

5. A strong stimulus produces a stronger action potential than does a weak stimulus.

6. Thick nerve fibers conduct impulses faster than thin ones do.

7. Dendrites do not secrete neurotransmitters.

8. A single EPSP is not enough to make a postsynaptic neuron fire.

9. cAMP acts by opening Na^+ gates in the plasma membrane.

10. The total effect of all EPSP's and IPSP's at one location is called summation.

D. Multiple Choice

1. The ions most permeable in neuron plasma membranes are: (a) potassium; (b) calcium; (c) sodium; (d) anions; (e) phosphate.

2. The central cavities of the brain and spinal cord are lined by: (a) neurons; (b) ependymal cells; (c) microglia; (d) Schwann cells; (e) astrocytes.

3. Which of the following is *not* characteristic of an action potential? (a) It is decremental. (b) It has the same voltage regardless of stimulus strength. (c) It is self-propagating. (d) It is irreversible. (e) It can occur only where there are voltage-gated ion gates.

4. During the repolarization phase of an action potential: (a) Ca^{2+} enters the axon; (b) Na^+ gates are open; (c) K^+ gates are open; (d) Na^+ gates are beginning to close; (e) an EPSP is generated.

5. In a ___ circuit of neurons, one input signal causes long-term, repetitive output because a neuron late in the circuit restimulates a neuron earlier in the circuit. (a) serial; (b) diverging; (c) converging; (d) parallel after-discharge; (e) reverberating

6. The effect of GABA on a postsynaptic neuron is: (a) hyperpolarization; (b) repolarization; (c) all-or-none; (d) depolarization; (e) excitation.

7. The first thing to happen at a synaptic knob when a nerve impulse arrives is: (a) Ca^{2+} influx; (b) Na^+ influx; (c) exocytosis of synaptic vesicles; (d) reuptake of ACh; (e) ACh binding to a receptor.

8. If a neuron has one axon and many dendrites, it is called: (a) unipolar; (b) bipolar; (c) unilateral; (d) bilateral; (e) multipolar.

9. When a neuron is at its resting membrane potential: (a) K^+ is more concentrated in the ECF than in the ICF; (b) K^+ is more concentrated in the ICF than in the ECF; (c) the Na^+–K^+ pump ensures that these two ions have equal concentrations in the ECF and ICF; (d) Na^+ is more concentrated in the ICF than in the ECF; (e) Ca^{2+} is more concentrated in the ICF than in the ECF.

10. All are true about unipolar neurons *except:* (a) they have one axon; (b) some have receptive dendrites in the skin; (c) they take action potentials toward the spinal cord; (d) they begin in an embryo as bipolar neurons; (e) they are most often motor neurons.

11. The all-or-none law of neurons states that: (a) all neurons have the same resting membrane potential; (b) all neurons conduct signals at the same speed; (c) a synaptic knob either releases all of its ACh or none of it; (d) a neuron will fire at maximum voltage once it has reached threshold; (e) regardless of stimulus strength, all EPSPs have the same voltage.

12. The monoamines include all of the following neurotransmitters *except:* (a) GABA; (b) norepinephrine; (c) dopamine; (d) serotonin; (e) histamine.

13. In an excitatory adrenergic synapse: (a) NE binds to ligand-gated ion channels; (b) the postsynaptic neuron is hyperpolarized; (c) IPSPs occur; (d) a second messenger system, such as cAMP, is activated; (e) acetylcholine is removed from the synapse by reuptake rather than by enzymatic degradation.

14. The ability of one neuron to make it easier for another neuron to stimulate a target cell is:
(a) temporal summation; (b) spatial summation; (c) facilitation; (d) reverberation; (e) neural coding.

15. In conduction of an AP in a myelinated axon: (a) conduction at nodes of Ranvier is very rapid; (b) nerve signals can travel across the surface of the myelin sheath; (c) conduction in the internodes is decremental but very fast; (d) initiation doesn't usually depend on neurotransmitters; (e) the process is aided by synaptic facilitation.

16. Neuropeptides that act as analgesics are: (a) CCK; (b) monoamines; (c) enkephalins; (d) amino acids; (e) catecholamines.

17. In a neurosoma, there are numerous dark-staining areas, called Nissl bodies, that are: (a) neurofibrils; (b) lysosomes; (c) rough ER; (d) the Golgi complex; (e) the cytoskeleton.

18. Cells and organs that detect changes in the internal and external environment are called: (a) effectors; (b) receptors; (c) integrators; (d) efferents; (e) afferents.

19. Slow axonal transport is: (a) the method by which some viruses and toxins get from a nerve ending to the neurosoma; (b) the method by which nerve signals are transmitted in unmyelinated nerve fibers; (c) a method of moving enzymes and cytoskeletal elements to the distal end of a nerve fiber; (d) the method by which nerve signals are transmitted in myelinated nerve fibers; (e) another name for retrograde transport.

20. When a neuron is hyperpolarized, its membrane potential is most likely to be: (a) –72 mV; (b) –65 mV; (c) –55 mV; (d) 0 mV; (e) +35 mV.

21. Neurotransmitters in the synaptic cleft must be eliminated, or the postsynaptic cell will continue to be stimulated. This can be achieved by:
   1. diffusion of the NT away from receptors on the postsynaptic membrane.
   2. enzymatic digestion of the NT in synaptic cleft.
   3. absorption of the NT by the synaptic knob.
   4. increased frequency of stimulation by the presynaptic neuron.

(a) 1 & 3; (b) 2 & 4; (c) 1, 2, & 3; (d) 4 only; (e) all the above

22. Action potentials:
   1. can begin with either depolarization or hyperpolarization.
   2. require stimulation of voltage-gated channels on the trigger zone and axon.
   3. are reversible and can return to RMP if threshold is not reached.
   4. are saltatory in myelinated fibers.

(a) 1 & 3; (b) 2 & 4; (c) 1, 2, & 3; (d) 4 only; (e) all the above

23. The neurotransmitter acetylcholine:
   1. was the first neurotransmitter discovered.
   2. is excitatory in heart muscle.
   3. is used at neuromuscular junctions in somatic motor neurons.
   4. is the major inhibitory neurotransmitter in the CNS.

(a) 1 & 3; (b) 2 & 4; (c) 1, 2, & 3; (d) 4 only; (e) all the above

24.   Which of these is/are true?
      1.   In an action potential, almost all sodium and potassium ions move across the membrane.
      2.   Electrical synapses require gap junctions, and some are found in cardiac muscle.
      3.   Synapses formed in childhood tend to remain constant throughout life.
      4.   The buildup of β–amyloid protein in the brains of Alzheimer patients seems to trigger the other manifestations of the disease.

      (a) 1 & 3; (b) 2 & 4; (c) 1, 2, & 3; (d) 4 only; (e) all the above

25.   Which of these is/are true?
      1.   One side effect of multiple sclerosis is neurosis.
      2.   Microglia are macrophages essential in synaptic remodeling during nervous system development.
      3.   Alzheimer's and Parkinson's diseases are both neurotransmitter disorders.
      4.   The effect a specific NT has on a postsynaptic membrane is determined by the target tissue and the receptor type present.

      (a) 1 & 3; (b) 2 & 4; (c) 1, 2, & 3; (d) 4 only; (e) all the above

E. Word Origins

1.   In *neurilemma, neuri-* means "nerve."
2.   In *afferent, af-* means "after, later."
3.   In *afferent, fer-* means "iron."
4.   In *oligodendrocyte, dendro-* means "branch."
5.   In *astrocyte, astro-* means "sky, heavens."
6.   In *astrocyte, -cyte* means "cell."
7.   In *sclerosis, sclero-* means "hard."
8.   In *synapse, -aps* means "gap, void."
9.   In *retrograde, retro-* means "backward."
10. In *ionotropic, trop-* means "place."
11. soma-
12. arbor-
13. oligo-
14. saltare-
15. ef -

F. Which One Does Not Belong?

1.   (a) synaptic vesicle; (b) nucleus; (c) Nissl bodies; (d) Golgi bodies

2.   (a) bipolar; (b) unipolar; (c) anaxonic; (d) multipolar

3.   (a) spinal cord; (b) somatic motor; (c) visceral motor; (d) somatic sensory

4.   (a) oligodendrocyte; (b) microglia; (c) ependymal cells; (d) Schwann cells

5.   (a) all-or-none; (b) inhibitory; (c) decremental; (d) reversible

G. Figure Exercise

Answer the following questions about this diagram.

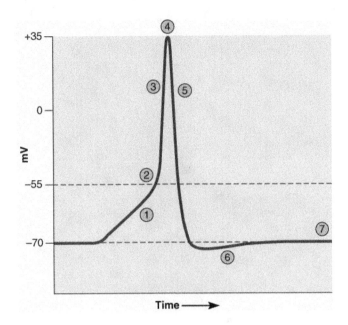

1. Period during which most K^+ is entering the neuron: (a) 1; (b) 2; (c) 3; (d) 7; (e) none of these

2. Time when more K^+ leaves the cell than Na^+ entered: (a) 2; (b) 3; (c) 4; (d) 5; (e) 6

3. Period when most voltage-gated Na^+ gates are open: (a) 1; (b) 2; (c) 3; (d) 5; (e) 6

4. Period of graded or local potential: (a) 1; (b) 3; (c) 4; (d) 6; (e) 7

5. Period when astrocytes remove excess K^+ from the ECF: (a) 2; (b) 4; (c) 5; (d) 6; (e) 7

6. Points contained in the absolute refractory period: (a) 4, 5, & 6; (b) 3, 4, & 5; (c) 5, 6, & 7; (d) 1, 2, & 3; (e) 1, 6, & 7

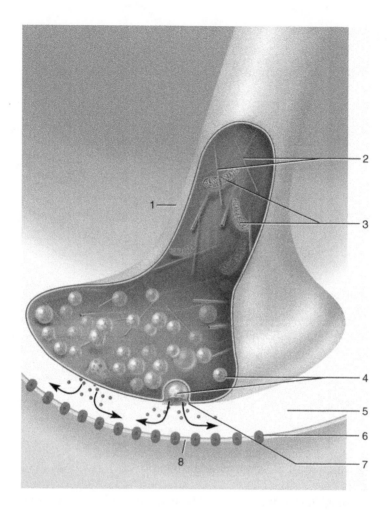

7. In order for the process shown in #7 to begin, what must occur *first?* (a) Mitochondria produce ATP. (b) NT vesicles diffuse to presynaptic membrane; (c) Action potentials on vesicle membranes move them to the presynaptic membrane. (d) Ca^{2+} from the ECF must enter the synaptic knob. (e) The receptors on #8 must open.

8. If this diagram were specific for an excitatory adrenergic synapse, what would be stored in #4? (a) NE; (b) GABA; (c) ACh; (d) serotonin; (e) cholecystokinin

9. Which of these is/are capable of an action potential? (a) 4 & 8; (b) 1 & 4; (c) 1 only; (d) 1, 4, & 8; (e) 1 & 8

10. If this were an excitatory cholinergic synapse, #6 would: (a) allow Na^+ in and K^+ out when bound by Ach; (b) allow Na^+ out and K^+ in when bound by ACh; (c) cause a chain reaction leading to cAMP production; (d) break down the NT soon after it binds; (e) inhibit further NT release.

13   The Spinal Cord, Spinal Nerves, and Somatic Reflexes

A. Short Answer

1. In the PNS, the cell bodies of neurons are concentrated in masses called ___.

2. The fibrous sheath around a nerve is called the ___.

3. All ___ nerves are mixed nerves.

4.   Sensory signals to the spinal cord enter it by way of its ___.

5. Above segment T6, the posterior column of the spinal cord is composed almost entirely of two ascending tracts of white matter, the ___ fasciculus and ___ fasciculus.

6. ___ is the crossing of nerve fibers from one side of the CNS to the other as they ascend from or descend to the brain or spinal cord.

7. White matter in the spinal cord is divided into columns that contain ___.

8. The axillary, radial, and ulnar nerves arise from the ___ plexus.

9. Of the motor nerve fibers leading to a skeletal muscle, the ___ neurons innervate the intrafusal fibers of the muscle spindles, and the ___ neurons innervate the extrafusal fibers that form most of the muscle.

10. A(n) ___ reflex arc is one with no interneuron and a minimum of synaptic delay. Such an arc is involved in the patellar tendon reflex, for example.

11. The meninges of the spinal cord include the outer ___, the middle ___, and the innermost ___.

12. The ___ ramus of a spinal nerve innervates the muscles, joints, and skin of the back at that spinal nerve level.

13. ___ in the spinal cord contains heavily myelinated nerve fibers.

14. Most cerebrospinal fluid around the spinal cord is found in the ___.

15. The spinal cord ends in the ___ at about the level of L1.

B. Matching

A. acetylcholine
B. dura mater
C. dermatome
D. anterior horn
E. muscle spindles
F. afferent

G. posterior horn
H. tract
I. plexus
J. tendon reflex
K. fascicle
L. alpha motor neuron

M. extrafusal fiber
N. efferent
O. endoneurium
P. decussate
Q. spinothalamic tract
R. funiculus

S. gamma motor neuron
T. crossed extension reflex
U. spinoreticular tract
V. flexor reflex
W. parasympathetic division
X. posterior root ganglion

1. Area of skin innervated by one spinal nerve
2. Fusiform proprioceptors in skeletal muscles that respond to stretch
3. Subdivision of a nerve consisting of multiple nerve fibers enclosed in a perineurium
4. Nerve fibers that conduct signals from the CNS to a muscle or gland
5. Employs a parallel after-discharge circuit to produce sustained muscle contraction
6. Adjusts the length of a muscle spindle
7. Innervate extrafusal fibers
8. Contains the neurosomas of afferent somatosensory neurons
9. Carries pain information from tissue injury
10. Inhibits muscle contraction to prevent muscle tearing

C. True or False

1. The lateral horn of spinal cord gray matter contains efferent fibers of somatic motor neurons.

2. Most nerves are sensory nerves.

3. When a very quick reflexive response is needed, it is likely to involve a spinal reflex arc with no interneurons.

4. Proprioceptors are sense organs that detect changes in position and movement of the body.

5. The origin and destination of a nerve tract are on the same side of the body.

6. Rabies is caused by a virus that affects the CNS.

7. Skeletal muscle contractions result from stimulation by gamma motor neurons.

8. A reflex is a rapid, stereotyped, involuntary response to a stimulus.

9. Testing somatic reflexes can aid in diagnosing diabetes mellitus.

10. The spinothalamic tract is a descending motor pathway that controls fine movements of the limbs.

D. Multiple Choice

1. Spinal nerves form all of the following plexuses *except:* (a) cervical; (b) brachial; (c) thoracic; (d) lumbar; (e) coccygeal.

2. A reflex arc always includes all of these *except:* (a) a motor neuron; (b) a receptor; (c) the brain; (d) an afferent neuron; (e) an effector.

3. Stubbing your left hallux, subsequently lifting your left foot and balancing on your right foot, is an example of: (a) the flexor reflex; (b) synaptic delay; (c) reciprocal inhibition; (d) the crossed extension reflex; (e) the tendon reflex.

4. Many nerves in the cauda equina innervate: (a) the hip region; (b) the thorax; (c) most of the skin; (d) the coccygeal region; (e) the lower extremities.

5. All somatosensory signals from the shoulders down enter the spinal cord by way of the: (a) anterior root; (b) posterior horn; (c) anterior horn; (d) central canal; (e) white matter.

6. All are functions of the spinal cord *except:* (a) conducts messages from the brain to effectors; (b) locomotion; (c) perception of sensations; (d) neural integration; (e) it is a reflex center.

7. The genitofemoral nerve, saphenous nerve, and obturator nerve arise from the: (a) ansa cervicalis; (b) brachial plexus; (c) lumbar plexus; (d) sacral plexus; (e) coccygeal plexus.

8. The primary afferent fibers in a stretch reflex come from: (a) the spinal cord; (b) flower-spray endings; (c) extrafusal fibers; (d) anulospiral endings; (e) the reticulospinal tract.

9. You get a paper cut and pull your hand back quickly. This reaction is a: (a) myotatic reflex; (b) monosynaptic reflex; (c) flexor reflex; (d) crossed-extension reflex; (e) tendon reflex.

10. The essential function of reciprocal inhibition in a somatic reflex is to: (a) produce a protective withdrawal response with minimum delay; (b) produce a prolonged muscle contraction; (c) prevent tearing of a muscle or tendon; (d) shift the body weight so you don't fall over; (e) prevent an antagonist from interfering with the contraction of a prime mover.

11. Given the anatomy of the spinal cord, epidural injections should be given: (a) below L5; (b) above L1; (c) at C1; (d) between L2 and L4; (e) between L5 and S4.

12. Proprioception signals from the lower extremities and lower trunk are carried mainly by the: (a) gracile fasciculus; (b) anterior spinocerebellar tract; (c) cuneate fasciculus; (d) corticospinal tract; (e) tectospinal tract.

13. The phrenic nerve, which innervates the diaphragm, is a branch of the ___ plexus. (a) cervical; (b) brachial; (c) thoracic; (d) lumbar; (e) sacral

14. The dermatome that encompasses the skin over the lateral malleolus is innervated by: (a) S1; (b) T12; (c) C4; (d) L4; (e) S4.

15. There are ___ pairs of spinal nerves. (a) 12; (b) 7; (c) 10; (d) 36; (e) 31

16. The descending tract that originates in the vestibular nucleus and controls limb muscles that maintain balance is the: (a) posterior spinocerebellar; (b) posterior spinothalamic; (c) lateral corticospinal; (d) cuneate fasciculus; (e) vestibulospinal.

17. The sympathetic chain of ganglia is connected to a spinal nerve by the: (a) posterior ramus; (b) meningeal branch; (c) segmental branch; (d) anterior root; (e) communicating rami.

18. In regions where the anterior rami of the spinal nerves do not form nerve plexuses, they give rise to: (a) intercostal nerves; (b) cranial nerves; (c) phrenic nerves; (d) nerves of the upper and lower extremities; (e) splanchnic nerves.

19. Nerve pain associated with pressure on spinal nerves is: (a) paresthesia; (b) neuralgia; (c) neuropathy; (d) paralysis; (e) sclerosis.

20. Ascending sensory pathways usually require ___ neurons, while descending motor pathways typically require ___ neurons. (a) three, two; (b) two, four; (c) three, four; (d) three, three; (e) four, three

21. Components of a muscle spindle include:
    1. anulospiral endings that respond to the beginning of muscle stimulation.
    2. nuclear bag fibers.
    3. afferent fibers sensitive to sudden movements.
    4. junctional folds on alpha motor neurons.

    (a) 1 & 3; (b) 2 & 4; (c) 1, 2, & 3; (d) 4 only; (e) all the above

22. Which of these is/are true about the brachial plexus?
    1. It forms the long thoracic nerves.
    2. Its cords in the armpits can be subject to "crutch" paralysis.
    3. It includes roots, trunks, and anterior and posterior divisions.
    4. It forms from posterior rami of C4–T2.

    (a) 1 & 3; (b) 2 & 4; (c) 1, 2, & 3; (d) 4 only; (e) all the above

23. Which of these is/are true?
    1. Reflexes that travel over unmyelinated fibers are very rapid.
    2. Reflexes occur only in somatic motor pathways.
    3. Gray matter in the spinal cord is divided into tracts.
    4. The tectospinal tract is involved with reflex movements of the head in response to visual and auditory stimuli.

    (a) 1 & 3; (b) 2 & 4; (c) 1, 2, & 3; (d) 4 only; (e) all the above

24. Which of these is/are true about dysfunctions of the nervous system?
    1. Shingles is caused by the chickenpox virus, which remains in posterior root ganglia of spinal nerves after infection.
    2. ALS is caused by toxic levels of the neurotransmitter glutamate.
    3. Poliomyelitis causes paralysis due to destruction of motor neurons.
    4. Spina bifida can be prevented if women of childbearing age eat plenty of green leafy vegetables, black beans, and/or lentils some months before getting pregnant.

    (a) 1 & 3; (b) 2 & 4; (c) 1, 2, & 3; (d) 4 only; (e) all the above

25. The sciatic nerve:
   1. arises from the sacral plexus.
   2. gives rise to the tibial and common fibular nerves.
   3. innervates muscles of the lower extremity.
   4. is a mixed nerve carrying somatosensory fibers as well as somatic motor fibers.

   (a) 1 & 3; (b) 2 & 4; (c) 1, 2, & 3; (d) 4 only; (e) all the above

E. Word Origins

   1.   In *ganglion, gangli-* means "tangled."
   2.   In *neuralgia, -algia* means "pain."
   3.   In *posterior ramus, ramus* means "branch."
   4.   In *somatosensory, somato-* means "body."
   5.   In *dermatome, -tom* means "cut apart."
   6.   In *intrafusal, intra-* means "within."
   7.   In *myotatic, -tatic* means "contraction."
   8.   In *contralateral, contra-* means "opposite."
   9.   In *baroreceptor, baro-* means "deep."
   10. In *sympathetic, path-* means "route."
   11. -oid
   12. dura
   13. polio
   14. proprio-
   15. bifid-

F.   Which One Does Not Belong?

   1. (a) posterior root; (b) lateral horn; (c) anterior root; (d) spinal nerve

   2. (a) rabies; (b) shingles; (c) poliomyelitis; (d) paresthesia

   3. (a) musculocutaneous nerve; (b) ulnar nerve; (c) ansa cervicalis; (d) axillary nerve

   4. (a) tectospinal; (b) spinocerebellar; (c) gracile fasciculus; (d) spinothalamic

   5. (a) poliomyelitis; (b) Guillain-Barré syndrome; (c) rabies; (d) Lou Gehrig' disease

G. Figure Exercise

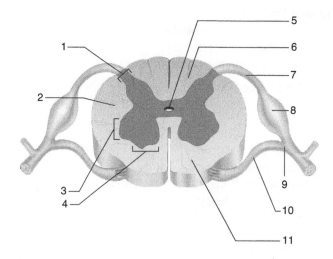

Identify the correct part of the figure.

1. Contains CSF at some time during life
   (a) 3; (b) 4; (c) 5; (d) 8; (e) 9

2. Contains neurosomas of somatic motor neurons.
   (a) 1; (b) 2; (c) 3; (d) 4; (e) 8

3. Contains neurosomas of ANS neurons in thoracic and lumbar regions
   (a) 1; (b) 2; (c) 3; (d) 4; (e) 8

4. House somatic motor and visceral motor fibers at some cord levels
   (a) 3 & 4; (b) 4 & 7; (c) 7 & 9; (d) 3 & 10; (e) 9 & 10

5. Contains neurosomas of sensory neurons
   (a) 3; (b) 6; (c) 8; (d) 9; (e) 11

6. Contain afferent fibers
   (a) 2, 6, 7, & 9; (b) 1, 6, 10, & 11; (c) 2, 6, & 11; (d) 1, 3, & 4; (e) 8, 9, 10, & 11

7. Does not contain myelinated nerve fibers
   (a) 2; (b) 3; (c) 11; (d) 6; (e) 10

8. Location of anterior spinocerebellar tract
   (a) 2; (b) 3; (c) 6; (d) 10; (e) 11

9. Contains fibers that aid in limb and trunk proprioception and movement below T6
   (a) 2; (b) 3; (c) 6; (d) 10; (e) 11

10. Contains fascicles
    (a) 1; (b) 4; (c) 8; (d) 9; (e) 10

14   The Brain and Cranial Nerves

A. Short Answer

1. The cell bodies of CNS neurons are concentrated in tissue called ___, found at the center of the spinal cord and on the surface of the cerebrum, among other places.

2. The CNS is covered by three fibrous membranes called the ___.

3. The falx cerebri and tentorium cerebelli that separate brain regions are part of the ___.

4. Many substances in the blood are unable to get into the brain because of the ___ which involves tight junctions between capillary endothelial cells.

5. Muscles of facial expression are controlled mainly by the ___ nerve.

6. The part of the limbic system that lies superior and rostral to the corpus callosum is the ___.

7. The longest cranial nerve, the ___, innervates viscera of the thoracic and abdominal cavities.

8. The ___ is the timekeeping center of the brain.

9. Degeneration of the ___ causes Parkinson disease.

10. Most commissural tracts of the cerebrum pass between the right and left hemispheres through a C-shaped bundle called the ___.

11. The ___ lobe of the cerebrum is concerned mainly with vision.

12. Brain neuropeptides called ___ stimulate wakefulness and increase metabolic rate.

13. Many emotional responses and perceptions of reward and punishment reside in the ___, a ring of structures on the medial border of the temporal lobe.

14. When you engage in intellectual activity, such as taking an exam, your electroencephalogram is dominated by ___ waves.

15. Each of the brain's areas of primary sensory cortex is surrounded by a(n) ___, where new sensory information is related to previous experience and thus interpreted or identified.

B. Matching

A. rostral	G. midbrain	M. stage 4	S. medulla oblongata
B. temporal lobe	H. motor speech	N. cerebellum	T. corpus callosum
C. thalamus	I. cauda equina	O. categorical hemisphere	U. ependyma
D. REM	J. association areas	P. fibroblasts	V. precentral gyrus
E. pons	K. postcentral gyrus	Q. hypothalamus	W. corpora quadrigemina
F. frontal lobe	L. caudal	R. reading	X. lateralization

1. Cell of the arachnoid meninx that suspend the brain in the skull
2. Integrates sensory input and relays it to the appropriate area of the cerebrum
3. Sleep stage in which most dreaming occurs
4. Explains specialization of functions in different brain hemispheres
5. Location of primary motor cortex
6. Site of cardiac, vasomotor, respiratory, and swallowing centers
7. Concerned with motor coordination and muscle tone
8. Inferior part of diencephalon forming the walls and floor of the third ventricle
9. Site of voluntary control of the skeletal muscles
10. Function of the Broca area

C. True or False

1. More cerebral cortex is dedicated to control of the hand than to control of all the muscles from the shoulder to the hip.

2. The optic nerve, among others, controls eyeball movement.

3. A 78-year-old man has suffered a stroke. Since then, he has been able to walk but he tends to lose his balance, feel dizzy, and sometimes fall down. We would suspect the stroke damaged his cerebellum more than we would suspect damage to the primary motor cortex.

4. Falling asleep and waking up are controlled mainly by the reticular formation of the brain stem and by the hypothalamus.

5. The wrinkles and creases in the cerebral cortex are called neural folds and neural grooves.

6. Cerebrospinal fluid is reabsorbed into the bloodstream by capillaries of the choroid plexuses.

7. In most people, mathematical calculations are carried out in the left cerebral hemisphere, while recognition of people's faces resides in the right.

8. The lumen of the embryonic neural tube gives rise to the central canal in the spinal cord and the brain ventricles.

9. The cerebral cortex can be recognized histologically by its Purkinje cells.

10. EEGs are recordings of the action potentials of nerve fibers in the cerebral cortex.

D. Multiple Choice

1. The cerebrum is derived from the ___ of the embryo. (a) telencephalon; (b) myelencephalon; (c) diencephalon; (d) metencephalon; (e) mesencephalon

2. Cerebrospinal fluid is secreted by the: (a) meninges; (b) cerebral aqueduct; (c) arachnoid villi; (d) circumventricular organs; (e) choroid plexuses and ependymal cells.

3. The nerve most likely to be damaged by a fracture of the ethmoid bone is the: (a) olfactory; (b) trochlear; (c) mandibular; (d) glossopharyngeal; (e) vestibulocochlear.

4. All of these statements are true about cerebrospinal fluid *except* that: (a) it forms the blood–brain barrier; (b) it cushions the brain from contact with the skull; (c) it flows continually in and around the CNS; (d) It is produced in all four ventricles. (e) It provides chemical stability for the brain.

5. The primary motor cortex of the right cerebral hemisphere controls skeletal muscles on the left side of the body because its nerve fibers decussate: (a) through the corpus callosum; (b) in the thalamus; (c) in the pyramids of the medulla oblongata; (d) in the reticulospinal tracts; (e) in the gray commissure of the spinal cord.

6. All sensory information to the cerebral cortex passes through the ___ before reaching the cortex. (a) anterior horns of the spinal cord; (b) posterior horns of the spinal cord; (c) posterior roots of the spinal nerves; (d) pons; (e) thalamus

7. Arbor vitae and Purkinje cells are characteristic of the histology of the: (a) spinal cord; (b) medulla oblongata; (c) cerebellum; (d) hypothalamus; (e) cerebral cortex.

8. The ___ control(s) the tracking movements in which the eyes turn to follow a moving object. (a) reticular formation; (b) superior colliculi of the midbrain; (c) olivary nucleus of the medulla oblongata; (d) occipital lobe of the cerebrum; (e) lateral geniculate body of the thalamus

9. All of these are functions of the hypothalamus *except*: (a) thermoregulation; (b) hormone production; (c) memory; (d) problem solving; (e) maintenance of circadian rhythms.

10. The ___ provide(s) feedback to the cerebral cortex that helps to coordinate the actions of the skeletal muscles. (a) basal nuclei; (b) limbic system; (c) tectum; (d) tegmentum; (e) nucleus cuneatus

11. The ___ nucleus produces a hormone useful in maintaining water balance. (a) dorsomedial; (b) arcuate; (c) mammillary; (d) preoptic; (e) supraoptic

12. The conscious awareness of smell is attributed to the: (a) temporal lobes; (b) inferior part of the frontal lobe; (c) primary sensory cortex; (d) limbic system; (e) hypothalamus.

13. The categorical hemisphere (in most people, the left) is probably more effective than the representational hemisphere for all of the following functions *except:* (a) speaking; (b) remembering scientific vocabulary; (c) remembering anatomical illustrations; (d) doing mathematical calculations; (e) understanding a lecture.

14. An example of procedural memory is remembering: (a) facts for an exam; (b) a person's name when you hear his/her voice; (c) your telephone number; (d) how to type or keyboard; (e) someone's birthday.

15. The only lobe of the cerebrum that cannot be seen from the surface is the: (a) frontal lobe; (b) temporal lobe; (c) parietal lobe; (d) occipital lobe; (e) insula.

16. The region of the cerebrum most concerned with hearing is the: (a) orbitofrontal cortex; (b) temporal lobe; (c) insula; (d) parietal lobe, within the longitudinal fissure; (e) occipital lobe, immediately posterior to the parieto-occipital sulcus.

17. The neural tube develops from the embryonic: (a) endoderm; (b) mesoderm; (c) ectoderm; (d) telencephalon; (e) myelencephalon.

18. A person who demonstrates musical and artistic talent probably has a well-developed: (a) representational hemisphere; (b) corpus callosum; (c) limbic system; (d) primary sensory cortex; (e) insula.

19. Neural crest cells in an embryo give rise to all *except:* (a) the dura mater; (b) Schwann cells; (c) ANS nerves; (d) ganglia of the PNS; (e) some endocrine structures.

20. A cranial nerve that is purely sensory (not mixed) is: (a) I; (b) III; (c) IV; (d) IX; (e) XII.

21. The reticular formation:
    1. contains neurosomas of efferent fibers that modulate pain.
    2. plays a central role in the control of states of consciousness.
    3. aids in adjusting skeletal muscle contractions that maintain balance and posture.
    4. includes the vasomotor and cardiac centers.

    (a) 1 & 3; (b) 2 & 4; (c) 1, 2, & 3; (d) 4 only; (e) all the above

22. Which of these is/are true?
    1. The Wernicke area, normally in the left hemisphere, allows recognition of written and spoken language.
    2. Meningitis can be caused by either viruses or bacteria.
    3. The affective language area is on the opposite side of the brain from the Wernicke area.
    4. Blocking afferent fibers in spinal cord white matter can lead to hydrocephalus.

    (a) 1 & 3; (b) 2 & 4; (c) 1, 2, & 3; (d) 4 only; (e) all the above

23. Which of these is/are true about the CNS?
    1. The hippocampus stores long-term memory.
    2. Control over the expression of emotions is located in the prefrontal cortex.
    3. The substantia nigra is a sensory relay center for memory.
    4. Forgetting is as important to mind function as is remembering.

    (a) 1 & 3; (b) 2 & 4; (c) 1, 2, & 3; (d) 4 only; (e) all the above

24. Which of these is/are true?
    1. The three meninges all have two layers: the periosteal and meningeal layers.
    2. Dural sinuses between the two layers of the dura mater contain CSF.
    3. The cerebral aqueduct connects the fourth ventricle with the central canal of the spinal cord.
    4. The blood–brain barrier is formed in brain capillaries by tight junctions whose development is stimulated by astrocyte perivascular feet.
    (a) 1 & 3; (b) 2 & 4; (c) 1, 2, & 3; (d) 4 only; (e) all the above

25. Which of these is/are true?
    1. Information shared between the right and left halves of the cerebrum is accomplished by commissures.
    2. Most neural integration takes place in gray matter of the midbrain.
    3. The amygdala and hippocampus are associated with emotion and memory.
    4. Temporal lobe lesions can lead to prosopagnosia, the inability to recognize familiar objects.

    (a) 1 & 3; (b) 2 & 4; (c) 1, 2, & 3; (d) 4 only; (e) all the above

E. Word Origins

1. In *cerebellum, -ellum* means "lower."
2. In *gyrus, gyr-* means "twist, turn."
3. In *corpus callosum, callos-* means "tough."
4. In *prosencephalon, encephal-* means "skull."
5. In *arachnoid, arachn-* means "spider."
6. In *hydrocephalus, cephal-* means "head."
7. In *decussation, decuss-* means "mixed up."
8. In *ipsilateral, ipsi-* means "same."
9. In *cerebral cortex, cortex* means "gray."
10. In *cingulate, cingul-* means "pain."
11. tele-
12. myel-
13. verm-
14. lenti-
15. -gnos

F. Which One Does Not Belong?

1. (a) cardiac center; (b) thermoregulation center; (c) respiratory center; (d) vasomotor center

2. (a) substantia nigra; (b) tectum; (c) corpora quadrigemina; (d) pons

3. (a) somatosensory area; (b) Wernicke area; (c) Broca area; (d) affective language area

4. (a) tasting peppermint ice cream; (b) hearing an opera; (c) playing the piano; (d) seeing a sunrise

5. (a) using your imagination; (b) solving a problem in calculus; (c) recognizing subtle differences in shades of green; (d) painting the Sistine Chapel

G. Cranial Nerve Matching

I.   Match the CN with the foramen it passes through in the skull. Some answers can be used more than once.

A. cribriform plate, ethmoid
B. foramen ovale, sphenoid
C. internal acoustic meatus, temporal bone
D. optic canal, sphenoid
E. superior orbital fissure, sphenoid

F. foramen rotundum, sphenoid
G. hypoglossal canal, occipital bone
H. jugular foramen, occipital and temporal bones
I. stylomastoid foramen, temporal bone

1. Olfactory
2. Optic
3. Oculomotor
4. Trochlear
5. Ophthalmic division of trigeminal
6. Maxillary division of trigeminal
7. Mandibular division of trigeminal

8. Abducens
9. Facial
10. Vestibulocochlear
11. Glossopharyngeal
12. Vagus
13. Accessory
14. Hypoglossal

II. Match the CN with its function. Some answers can be used more than once.

A. I
B. II
C. III
D. IV

E. V_1
F. V_2
G. V_3
H. VI

I. VII
J. VIII
K. IX
L. X

M. XI
N. XII

1. Carries ANS fibers to sublingual salivary glands
2. Carries ANS motor fibers to stomach
3. Carries somatic motor fibers to stapedius muscle
4. Innervates lateral rectus muscle
5. Innervates muscles of facial expression
6. Innervates muscles of mastication
7. Innervates stylopharyngeal muscles
8. Innervates trapezius muscle (although it's not a true CN)
9. Motor to extrinsic tongue muscles
10. Motor to outer hair cells of cochlea
11. Motor to parotid gland
12. Motor to sternocleidomastoid muscle
13. Movements of tongue for speech
14. Opening of eyelid
15. Pupil constriction
16. Sensory fibers for hearing and equilibrium
17. Sensory from abdominal viscera
18. Sensory from dura mater
19. Sensory from lower face
20. Sensory from pharynx
21. Sensory from upper teeth and gums
22. Smell
23. Stimulates superior oblique muscle
24. Taste buds on posterior one-third of tongue
25. Vision

15 The Autonomic Nervous System and Visceral Reflexes

A. Short Answer

1. The autonomic nervous system has ___ and ___ divisions.

2. Two nerve branches called ___ connect each sympathetic ganglion to a spinal nerve.

3. Efferent pathways of the sympathetic nervous system have short ___ fibers and long ___ fibers.

4. The two types of receptors for acetylcholine are ___ receptors, found at neuromuscular junctions, and ___ receptors, found on most parasympathetic target cells.

5. The cells of the adrenal medulla behave like endocrine gland cells but are actually modified ___.

6. The parasympathetic nervous system issues its fibers through certain ___ and sacral nerves.

7. A nerve fiber that secretes acetylcholine or a receptor that binds ACh is said to be ___.

8. Many organs receive both sympathetic and parasympathetic fibers and, thus, are said to have ___.

9. Some organs receive only sympathetic fibers that have a steady background firing rate called ___.

10. The ___ of the brain controls many fundamental homeostatic mechanisms and exerts its control through autonomic efferent fibers.

11. Whether a neurotransmitter has ___ or ___ effects depends on the type of receptor to which it binds.

12. In the sympathetic division, most preganglionic fibers synapse with postganglionic fibers in a(n) ___ of ganglia alongside the spinal column.

13. The ___ is called the craniosacral division of the ANS.

14. ___ fibers in the ANS have their neurosomas in the CNS.

15. The ___ nerves originate mainly from spinal nerves T5–T12 in the sympathetic division and lead to collateral ganglia.

B. Matching

A. sympathetic tone
B. vagus nerve
C. intramural
D. norepinephrine
E. preganglionic
F. neuropeptides

G. lateral horn
H. parasympathetics
I. acetylcholine
J. terminal/intramural
K. parasympathetic tone
L. acetylcholinesterase

M. splanchnic nerve
N. celiac
O. pons
P. glossopharyngeal nerve
Q. adrenal cortex
R. somatic motor

S. monoamine oxidase
T. norepinephrine
U. sympathetics
V. spinal nerve
W. neurotransmitters
X. sympathetic trunk

1. Visceral motor fibers originate here
2. Neurotransmitter of all preganglionic fibers in the ANS
3. Maintains resting heart rate
4. Has only one neuron between CNS and effector
5. Has short, unmyelinated postganglionic fibers
6. Modulate the effects of ACh and NE in the ANS
7. Breaks down norepinephrine
8. Carries fibers controlling salivation and blood pressure
9. Released by postganglionic adrenergic fibers in the sympathetic division
10. Ganglion in sympathetic division

C. True or False

1. Sympathetic fibers that innervate sweat glands are cholinergic.

2. When norepinephrine binds alpha receptors in blood vessels, the result is vasoconstriction.

3. Sympathomimetics can stimulate norepinephrine release.

4. Vasodilation of cardiac and skeletal muscle blood vessels occurs as a result of acetylcholine binding to smooth muscle receptors.

5. Most organs in the body are innervated by either the sympathetic division or the parasympathetic division of the ANS.

6. Caffeine may make us more alert by blocking the actions of GABA, which makes us sleepy.

7. Nicotinic cholinergic receptors act by opening ligand-gated channels.

8. The heart requires ANS innervation in order to function.

9. Caffeine effectively blocks the neuromodulator adenosine; this allows more ACh secretion and we feel more alert.

10. Both the ANS and somatic motor divisions contain unmyelinated axons.

D. Multiple Choice

1. Some nuclei of the cranial nerves that carry ANS fibers are located in the: (a) thalamus; (b) cerebellum; (c) cerebral cortex; (d) medulla oblongata; (e) spinal cord.

2. Some cold medicines cause dilation of the bronchioles by binding alpha receptors to reduce swelling in nasal mucosa. These drugs are: (a) parasympathomimetics; (b) sympathomimetics; (c) used to treat Raynaud disease; (d) sympatholytics; (e) norepinephrine.

3. Ganglia in the sympathetic division include all *except:* (a) collateral; (b) superior mesenteric; (c) celiac; (d) ciliary; (e) inferior mesenteric.

4. Parasympathetic fibers to some salivary glands are carried over the: (a) facial nerve; (b) submandibular nerve; (c) splanchnic nerve; (d) oculomotor nerve; (e) vagus nerve.

5. Sympathetic effects include all *except:* (a) increased heart rate; (b) increased blood flow to the skin; (c) increased blood flow to cardiac muscle; (d) increased blood glucose; (e) increased alertness.

6. A cranial nerve involved in the baroreceptor reflex when blood pressure increases is the: (a) vagus; (b) hypoglossal; (c) oculomotor; (d) trigeminal; (e) ansa cervicalis.

7. Parasympatholytic drugs: (a) stimulate ACh release; (b) stimulate NE release; (c) inhibit ACh release; (d) bind alpha receptors; (e) bind adrenergic receptors.

8. During embryonic development, neural crest cells migrate to the colon and form the: (a) celiac ganglion; (b) enteric nervous system; (c) parasympathetic nervous system; (d) sympathetic trunk; (e) splanchnic nerves.

9. Most ganglia of the parasympathetic division are found: (a) in the thoracic cavity; (b) in or near the target organ; (c) in the brain; (d) in the head; (e) associated with cranial nerves III–XII.

10. Splanchnic nerves carry postganglionic fibers to all *except* the: (a) urinary bladder; (b) stomach; (c) kidneys; (d) spleen; (e) proximal colon.

11. Norepinephrine is secreted by most: (a) sympathetic preganglionic fibers; (b) sympathetic postganglionic fibers; (c) parasympathetic preganglionic fibers; (d) parasympathetic postganglionic fibers; (e) somatic motor fibers.

12. Because of ___, the sympathetic division has more widespread effects than the parasympathetic division. (a) greater neural divergence; (b) the adrenal medulla; (c) norepinephrine; (d) acetylcholine; (e) adrenergic receptors

13. The sympathetic division of the ANS produces all of the following effects *except:* (a) glycogen breakdown; (b) labor contractions; (c) bronchodilation; (d) gastric secretion; (e) dilation of the pupils.

14. The cholinergic effects of the parasympathetic division are exerted on most target cells through: (a) alpha receptors; (b) beta receptors; (c) muscarinic receptors; (d) nicotinic receptors; (e) cholinergic receptors.

15. Most parasympathetic preganglionic fibers travel through the: (a) splanchnic nerves; (b) pelvic nerves; (c) glossopharyngeal nerves; (d) facial nerves; (e) vagus nerves.

16. All are true about the somatic motor division *except* that it: (a) innervates skeletal muscle; (b) uses ACh as its neurotransmitter; (c) is always excitatory; (d) innervates sweat glands in the skin; (e) requires cholinergic receptors.

17. Fibers in the gray communicating rami: (a) are wrapped with oligodendrocytes; (b) are connecting points for parasympathetic nerves and the spinal cord; (c) are unmyelinated; (d) are heavily myelinated by Schwann cells; (e) extend from the lumbar spine to organs they innervate.

18. Spinal nerves T1–L2 contain: (a) postganglionic parasympathetic fibers; (b) preganglionic parasympathetic fibers; (c) neurosoma of sympathetic neurons; (d) preganglionic sympathetic fibers; (e) unmyelinated fibers.

19. ___ is/are common in the sympathetic division but *not* in the parasympathetic division of the ANS. (a) Adrenergic receptors; (b) Myelinated fibers; (c) Acetylcholine; (d) Intramural ganglia; (e) A two-neuron chain from spinal cord to target organ

20. Dual innervation is absent in all of the following *except:* (a) sweat glands; (b) the heart; (c) piloerector muscle; (d) the adrenal medulla; (e) blood vessels of skeletal muscle.

21. Cholinergic fibers include all:
    1. preganglionic fibers in the parasympathetic division.
    2. postganglionic fibers in the parasympathetic division.
    3. preganglionic fibers in the sympathetic division.
    4. postganglionic fibers in the sympathetic division.

    (a) 1 & 3; (b) 2 & 4; (c) 1, 2, & 3; (d) 4 only; (e) all the above

22. The oculomotor nerve:
    1. carries motor fibers to muscles of the eye for eyeball movement.
    2. contains preganglionic sympathetic fibers.
    3. carries motor fibers to the pupil and lens of the eye.
    4. carries sensory fibers from the retina of the eye.

    (a) 1 & 3; (b) 2 & 4; (c) 1, 2, & 3; (d) 4 only; (e) all the above

23. Which of these is/are innervated by the parasympathetic division of the ANS?
    1. cardiac muscle
    2. sweat glands
    3. digestive system
    4. pancreas

    (a) 1 & 3; (b) 2 & 4; (c) 1, 2, & 3; (d) 4 only; (e) all the above

24. Which of these is/are true?
    1. Vasomotor tone is produced by the sympathetic division and maintains normal blood pressure.
    2. The limbic system is a connection between the ANS and cerebral cortex.
    3. Salivation is an example of cooperation between the two divisions of the ANS.
    4. Dual innervation of some organs is unequal between the parasympathetic and sympathetic divisions.

    (a) 1 & 3; (b) 2 & 4; (c) 1, 2, & 3; (d) 4 only; (e) all the above

25. Which of these is/are true?
    1. Neural convergence is most common in the parasympathetic division.
    2. The enteric nervous system innervates smooth muscle and glands.
    3. Sympathetic fibers travel along the facial nerve to tear glands.
    4. Spinal nerves carry sympathetic fibers to targets in the body wall.

    (a) 1 & 3; (b) 2 & 4; (c) 1, 2, & 3; (d) 4 only; (e) all the above

E. Word Origins

    1. In *baroreflex, baro-* means "pressure."
    2. In *paravertebral, para-* means "next to."
    3. In *autonomic, nom-* means "name."
    4. In *sympathetic, path-* means "way."
    5. In *splanchnic, splanchn-* means "viscera."
    6. In *adrenal, ren-* means "kidney."
    7. In *intramural, intra-* means "within."
    8. In *adrenal, ad-* means "above."
    9. In *enteric, enter-* means "within."
   10. In *sympatholytics, -lytic* means "static."
   11. auto-
   12. ramus
   13. sym-
   14. mur-
   15. mimet-

F. Which One Does Not Belong?

    1. (a) acetylcholine; (b) muscarinic receptor; (c) norepinephrine; (d) nicotinic receptor

    2. (a) skeletal muscle; (b) cardiac muscle; (c) involuntary; (d) smooth muscle

    3. (a) paravertebral ganglia; (b) collateral ganglia; (c) pterygopalatine ganglion; (d) superior mesenteric ganglion

    4. (a) cardiac plexus; (b) abdominal aortic plexus; (c) esophageal plexus; (d) pulmonary plexus

    5. (a) lacrimal gland secretion; (b) sweat gland secretion; (c) piloerector muscle contraction; (d) urinary bladder wall contraction

G. Identification Exercise

Label the functions listed 1–25 with either the parasympathetic (P) or sympathetic (S) division.

1. Stimulation of blood clotting
2. Decreased force and rate of heartbeat
3. Tear secretion
4. Close vision
5. Vasodilation of blood vessels in skeletal muscles
6. Increased gut motility
7. Vasodilation of most blood vessels
8. Insulin secretion
9. Penile erection
10. Increased secretion from gastric glands
11. Pupil dilation
12. Far vision
13. Glycogen breakdown
14. Contraction of smooth muscle in bladder wall
15. Decreased urine output
16. Pupil constriction
17. Glycogen synthesis
18. Decreased gut motility
19. Contraction of internal urethral sphincter
20. Sweat gland stimulation
21. Adrenal medulla stimulation
22. Bronchiole constriction
23. Blushing
24. Increased fat breakdown
25. Decreased fat breakdown

16   Sense Organs

A. Short Answer

1. Sense organs that monitor muscle contractions and joint movements are called ___.

2. Of the two types of pain, which is transmitted by myelinated nerve fibers?

3. ___ papillae at the rear of the tongue contain about half of our taste buds.

4. Olfactory fibers from the posterior nasal cavity synapse with neurons in the ___ of the brain.

5. The pitch of a sound is determined by the ___ of vibration.

6. The auditory ossicle attached to the tympanic membrane is the ___.

7. The transducer for the sense of hearing is a coiled, fluid-filled duct called the ___.

8. The cupula is a gelatinous membrane that covers the hair cells of the ___.

9. All the organs that produce tears and drain them from the eye are collectively called the ___ apparatus.

10. The type of tongue papillae that degenerates in early childhood are the ___ papillae.

11. If you are walking slowly, and then speed up, the acceleration will be sensed by the ___ of the inner ear.

12. The ___ of the ear have V-shaped rows of stereocilia, and they tune the cochlea.

13. The space between the cornea and lens of the eye is filled with fluid called ___.

14. Rod cells are lacking from an area of the retina called the ___, and cone cells are smallest and most closely spaced there.

15. The absorption of light converts the *cis* isomer of ___ to an all-*trans* isomer.

B. Matching

A. fungiform papillae	G. crista ampullaris	M. optic disc	S. tectorial membrane
B. basilar membrane	H. thalamus	N. nasal bone	T. macula lutea
C. enkephalin	I. rods	O. referred pain	U. macula sacculi
D. ganglion	J. acetylcholine	P. bradykinin	V. glutamate
E. palatine bone	K. ethmoid	Q. fast pain	W. outer hair cells
F. filiform papillae	L. optic chiasma	R. cones	X. bipolar

1. Location where some optic nerve fibers from each eye cross over to the opposite side of the brain
2. Receptor cells for daylight and color vision
3. Released by rods in the dark
4. Bone penetrated by the olfactory nerve fascicles
5. Senses the texture of food
6. Supports hair cells of the cochlea
7. Results from convergence of neural pathways in the CNS
8. Cells that can detect fluctuations in light intensity
9. Senses rotation of the head
10. Endogenous peptide with powerful analgesic effects

C. True or False

1. Stem cells in the corneal epithelium provide a means for regeneration if the cornea is damaged.

2. The posterior chamber of the eye is filled with vitreous humor.

3. Each cone has three types of pigment specialized to absorb light in three different regions of the visible spectrum.

4. The auditory ossicles increase the frequency of vibrations from the tympanic membrane to the inner ear.

5. Sustained loud music can fracture stereocilia in the cochlea.

6. Substance P blocks the transmission of pain signals from peripheral nerves to the brain.

7. Tactile (Meissner) corpuscles endow the skin with its sense of light touch.

8. Not all taste buds are on the tongue.

9. The majority of cochlear hair cells do not generate the signals we hear.

10. The signals of hearing are generated as hair cell stereocilia are bent over by the back-and-forth movement of the endolymph.

D. Multiple Choice

1. The ear is classified as a: (a) chemoreceptor; (b) thermoreceptor; (c) nociceptor; (d) mechanoreceptor; (e) photoreceptor.

2. Skin temperature is sensed by: (a) end bulbs; (b) lamellar corpuscles; (c) free nerve endings; (d) peritrichial endings; (e) tactile discs.

3. The sensation of pain results partly from the stimulation of pain receptors by: (a) substance P; (b) enkephalins; (c) bradykinin; (d) endorphins; (e) glutamate.

4. Taste buds are absent from: (a) filiform papillae; (b) vallate papillae; (c) fungiform papillae; (d) lingual papillae; (e) foliate papillae.

5. Odors often produce emotional responses as olfactory signals reach the: (a) olfactory mucosa; (b) olfactory bulbs; (c) orbitofrontal cortex; (d) reticular formation; (e) amygdala.

6. All of the following structures vibrate in synchrony with a sound wave *except* the: (a) tympanic membrane; (b) stapes; (c) stapedius; (d) oval window; (e) basilar membrane.

7. Eye movements are controlled by extrinsic muscles, including all of the following *except* the: (a) superior rectus; (b) superior oblique; (c) inferior rectus; (d) lateral rectus; (e) lateral oblique.

8. After tears flow across the eye, they are taken up by openings called: (a) lacrimal puncta; (b) nasolacrimal ducts; (c) medial commissures; (d) lacrimal sacs; (e) conjunctivae.

9. The cochlea consists of a fleshy tube that spirals around a bony, screw-like core called the: (a) malleus; (b) malleolus; (c) modiolus; (d) mobius; (e) motorola.

10. The vascular layer of the eye includes all of the following *except:* (a) the cornea; (b) the iris; (c) the choroid; (d) chromatophores; (e) the ciliary body.

11. Outer hair cells of the cochlea have their microvilli embedded in the: (a) vestibular membrane; (b) tectorial membrane; (c) basilar membrane; (d) cupula; (e) vitreous body.

12. ___ cells of the retina are inhibited by glutamate in the dark but excited in the light when glutamate secretion stops. (a) Ganglion; (b) Rod; (c) Bipolar; (d) Amacrine; (e) Cone

13. Cells in the retina include all *except:* (a) ganglion cells; (b) ciliary cells; (c) rods; (d) amacrine cells; (e) horizontal cells.

14. As you focus on a nearby object while it moves closer to your face, all of the following processes occur *except* the: (a) pupil constricts; (b) medial rectus muscles contract; (c) lenses become thicker; (d) ciliary muscle relaxes; (e) eye shifts from the scotopic to photopic mode.

15. Otoliths add weight, and thus inertia, to the membranes of the: (a) cochlea; (b) saccule; (c) semicircular ducts; (d) ora serrata; (e) olfactory mucosa.

16. The modality of a stimulus is: (a) its location; (b) its intensity; (c) its type, such as vision versus hearing; (d) whether it is consciously perceived; (e) whether it is phasic or tonic.

17. Enkephalin blocks the release of ___ by afferent nerve endings in the spinal cord. (a) epinephrine; (b) bradykinin; (c) acetylcholine; (d) endorphin; (e) substance P

18. The olfactory tracts of the brain are composed of the axons of: (a) olfactory neurons; (b) mitral cells and tufted cells; (c) Purkinje cells; (d) pyramidal cells; (e) hair cells.

19. The ___ admits air to the middle-ear cavity to equalize pressure on both sides of the tympanic membrane. (a) auditory canal; (b) auditory tube; (c) oval window; (d) round window; (e) cochlear duct

20. In the transmission pathway for visual signals from receptor cells to the occipital lobe, action potentials occur in the: (a) rods and cones; (b) bipolar cells; (c) ganglion cells; (d) amacrine cells; (e) lateral geniculate body.

21. Which of these is/are true about afflictions of the sense organs?
    1. Astigmatism occurs if the eyeball is too short.
    2. Nerve deafness can be caused by destruction of hair cells.
    3. Presbyopia is a form of nearsightedness.
    4. Conductive deafness can result from otosclerosis.

    (a) 1 & 3; (b) 2 & 4; (c) 1, 2, & 3; (d) 4 only; (e) all the above

22. Which of these is/are true about the eye?
    1. Aqueous humor is secreted by the ciliary body.
    2. Bipolar cells of the retina are first-order neurons of the visual pathway.
    3. The second-order neurons of the visual pathway are ganglion cells.
    4. There are more nerve fibers in the optic nerve than there are rods and cones in the retina.

    (a) 1 & 3; (b) 2 & 4; (c) 1, 2, & 3; (d) 4 only; (e) all the above

23. Which of these is/are true about pain and its modulation?
    1. When dynorphins inhibit the release of substance P, no pain is perceived by the brain.
    2. Free nerve endings are nociceptors in epithelia and connective tissue.
    3. Bradykinin is a potent pain stimulus.
    4. In spinal gating, pain signals are allowed to reach the cerebellum via the posterior horns.

    (a) 1 & 3; (b) 2 & 4; (c) 1, 2, & 3; (d) 4 only; (e) all the above

24. Which of these is/are true?
    1. Amplitude of vibrations determines the pitch of sounds we hear.
    2. Secretions of apocrine sweat glands in women can stimulate synchronous menstrual cycles in other women.
    3. Rods and cones are the only neurons in the body directly exposed to the external environment.
    4. Olfactory cells are neurons whose cilia can bind odor molecules.

    (a) 1 & 3; (b) 2 & 4; (c) 1, 2, & 3; (d) 4 only; (e) all the above

25. Which of these is/are true about the mechanisms of hearing and balance?
    1. Potassium gradients in the endolymph provide the potential energy for hair cells to function properly.
    2. The tympanic reflex enhances the transfer of vibrations from the tympanic membrane to the round window.
    3. The stapedius and tensor tympani help coordinate speech with hearing.
    4. Vestibular nerve fibers terminate in the medulla oblongata and primary somesthetic cortex.

    (a) 1 & 3; (b) 2 & 4; (c) 1, 2, & 3; (d) 4 only; (e) all the above

E. Word Origins

1. In *nociceptor, noci-* means "pain."
2. In *analgesic, alges-* means "pain."
3. In *tympanic, tympan-* means "ear."
4. In *vestibulocochlear, cochle-* means "snail."
5. In *punctum, punct-* means "point."
6. In *binaural, -aur* means "sound."
7. In *otolith, oto-* means "ear."
8. In *otolith, -lith* means "jelly."
9. In *scotopic, scot-* means "light."
10. In *fovea centralis, fovea* means "patch."
11. foli-
12. -osis
13. macula
14. stereo-
15. –phore

F. Which One Does Not Belong?

1. (a) inner hair cells; (b) stereocilia; (c) tectorial membrane; (d) scala vestibuli

2. (a) nociceptor; (b) interoreceptor; (c) exteroreceptor; (d) proprioceptors

3. (a) basilar membrane; (b) tectorial membrane; (c) tympanic membrane; (d) otolithic membrane

4. (a) tactile corpuscle; (b) tactile disc; (c) lamellar corpuscles; (d) bulbous corpuscle

5. (a) myopia; (b) emmetropia; (c) presbyopia; (d) hyperopia

G. Figure Exercises

Match the statements with the structure in the diagrams. Some letters may be used more than once.

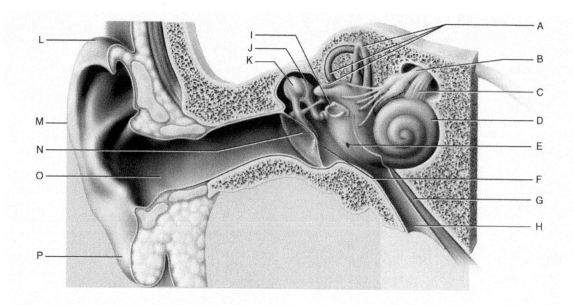

1. Vibrates when stimulated by sound waves
2. Contains scala vestibuli and cochlea
3. Attaches to oval window
4. Carries information about equilibrium to pons and cerebellum
5. Continuous with mastoid air cells
6. Contains ceruminous sebaceous glands
7. Allows equalization of pressure on both sides of tympanic membrane
8. Connected to sacculi and macula utriculi
9. Contains axons of bipolar neurons that terminate in pons

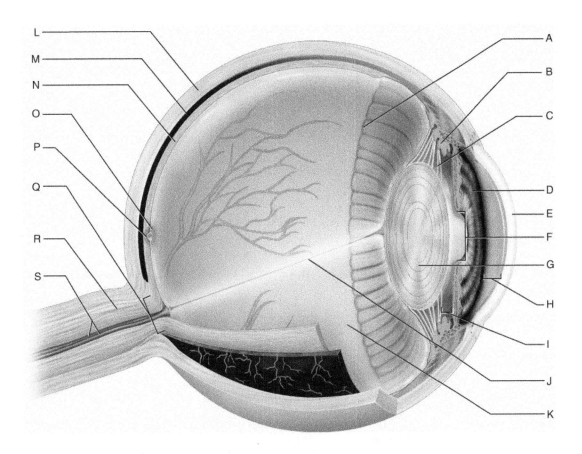

10. Allows eye to focus on far and near objects
11. Contains fovea centralis
12. Supports iris and lens and secretes aqueous humor
13. Place where optic nerve leaves fundus
14. Vascularized dense collagenous connective tissue layer
15. Area first to receive aqueous humor
16. Contains photoreceptor cells
17. Alters lens shape
18. Contains chromatophores and controls pupil diameter
19. Vascular pigmented layer behind retina
20. Contains smooth muscle and myoepithelial cells under ANS control

17 The Endocrine System

A. Short Answer

1. Some hormones are produced by nerve cells called ___.

2. Both PRL and TSH are released from the anterior pituitary when stimulated by ___.

3. Releasing and inhibiting hormones produced by the hypothalamus travel to the anterior pituitary by way of blood vessels called the ___.

4. Some endocrine glands elsewhere in the body suppress the secretion of hormones by the pituitary. This is called ___.

5. Hypersecretion of growth hormone causes ___ if it occurs in childhood or adolescence and ___ if it occurs in adulthood.

6. The metabolic rate is regulated by a large endocrine gland near the larynx called the ___.

7. Hormones of the adrenal cortex fall into three categories: sex steroids, ___, and ___.

8. The endocrine part of the pancreas consists of cell clusters called ___.

9. A group of hormones called ___ stimulate gluconeogenesis, protein and fat catabolism, and have anti-inflammatory effects on the body.

10. The interstitial cells of the testis secrete ___.

11. Thyroid hormone is hydrophobic, and it depends on a plasma protein, ___, to transport it in the blood.

12. Exocrine secretions have ___ effects, while endocrine secretions have ___ effects.

13. Polypeptide hormones cannot enter their target cells and, thus, depend on ___, such as cAMP and IP_3, to activate metabolic changes within them.

14. Two symptoms of diabetes mellitus are ___, which means glucose in the urine, and ___, which means an abnormally high level of glucose in the blood.

15. Because of the phenomenon of ___, a single-hormone molecule can ultimately cause a target cell to produce millions of molecules of a metabolic product.

B. Matching

A. kinase
B. phospholipase
C. prostaglandin
D. gastrin
E. Graves disease
F. renin

G. somatostatin
H. type1 DM
I. gonadotropin
J. progesterone
K. luteotropin
L. dopamine

M. ACTH
N. corticotropin
O. endemic goiter
P. insulin
Q. hypoglycemia
R. thyroxine

S. ghrelin
T. erythropoietin
U. pro-opiomelanocortin
V. lipoxygenase
W. triiodothyronine
X. myxedema

1. Inhibits GH secretion from adenohypophysis
2. Autoimmune disorder leading to increased heart rate and exopthalmos
3. Hormone that stimulates HCl secretion by the stomach
4. Enzyme that phosphorylates other enzymes
5. Formed when two diiodotyrosine residues are linked
6. Hormone of the kidneys and liver that increases red blood cell production
7. Caused by certain viruses acting on genetically susceptible people
8. Hormone of nutrient abundance
9. Any pituitary hormone that stimulates the ovaries or testes
10. Stimulates release of glucocorticoids from adrenal cortex

C. True or False

1. Hormones are secreted by bone and adipose tissue, the stomach, the placenta, and the heart.

2. Steroid hormones bind to receptors in the cytoplasm of their target cells.

3. A second messenger helps a polypeptide hormone such as oxytocin get through the plasma membrane of the target cell.

4. Hyperglycemic hormones include cortisol, glucagon, epinephrine, and growth hormone.

5. The neurohypophysis synthesizes oxytocin and antidiuretic hormone which are then transported by axoplasmic flow to secretion sites.

6. Hormones with nuclear receptors produce rapid responses in their target cells.

7. Some hormones oppose the effects of other hormones.

8. Insulin is secreted when blood glucose drops too low and helps raise it to a safe level.

9. Most cases of diabetes mellitus cannot be treated with insulin.

10. In the alarm reaction caused by stress, the secretion of corticotropin-releasing hormone precedes the secretion of glucocorticoids.

D. Multiple Choice

1. If hormones A and B have very little effect when each acts alone, but have a very large effect when they work together, they are said to have a(n) ___ effect on each other. (a) additive; (b) synergistic; (c) multiplicative; (d) positive feedback; (e) enzyme amplification

2. All are true about endocrine functions of the kidney *except:* (a) it secretes cholecalciferol; (b) it makes most of the body's EPO; (c) it produces an enzyme that converts angiotensinogen to angiotensin I; (d) it makes vitamin D_3; (e) a pathway begins here that ultimately results in aldosterone secretion from the adrenal cortex.

3. The anterior pituitary can be stimulated by: (a) corticotropin-releasing hormone; (b) corticotropic hormone; (c) interstitial cell–stimulating hormone; (d) thyroid-stimulating hormone; (e) somatomedin.

4. Blood calcium concentration is controlled mainly by a hormone from the: (a) anterior pituitary; (b) posterior pituitary; (c) adrenal cortex; (d) kidneys; (e) parathyroids.

5. ___ is referred to as the "salt-retaining hormone," and it helps correct low blood pressure. (a) ADH; (b) Renin; (c) Atrial natriuretic peptide; (d) Angiotensin II; (e) Aldosterone

6. In ___, the rapid, incomplete oxidation of fats gives rise to high levels of ketone bodies in the blood. (a) Cushing syndrome; (b) Addison disease; (c) atherosclerosis; (d) diabetes mellitus; (e) hypothyroidism

7. Antidiuretic hormone: (a) is synthesized in the pituitary; (b) stimulates the liver to release somatomedins; (c) reduces urine output; (d) targets bone tissue; (e) is a steroid.

8. Which of the following hormones binds to receptors on the plasma membrane of the target cell? (a) thyroid hormone; (b) aldosterone; (c) estrogen; (d) androgens; (e) prolactin

9. Which of the following hormones must travel through the hypophyseal portal system to carry out its function? (a) growth hormone; (b) gonadotropin-releasing hormone; (c) thyroxine; (d) luteinizing hormone; (e) cortisol

10. Insulin injections will not benefit people with type 2 diabetes mellitus because: (a) their alpha cells won't secrete insulin, anyway; (b) their beta cells antagonize the effect of the insulin; (c) their target cells lack receptors for insulin; (d) when insulin binds to the target cell it fails to activate the second messenger; (e) they cannot metabolize glucose even if the target cells do absorb it.

11. Transport proteins are necessary for ___ to travel in the bloodstream because of the hydrophobic nature of this hormone. (a) insulin; (b) oxytocin; (c) epinephrine; (d) estrogen; (e) antidiuretic hormone

12. Inositol triphosphate (IP$_3$) carries out its second-messenger role in hormone action by: (a) transporting the hormone in the blood; (b) helping the hormone get through the plasma membrane of the target cell; (c) opening voltage-gated channels in the target cell plasma membrane; (d) releasing calcium ions from smooth endoplasmic reticulum; (e) phosphorylating dormant cytoplasmic enzymes.

13. Chronic stress can make a person more susceptible to infection or cancer because it suppresses the immune system and stimulates the continual secretion of: (a) cortisol; (b) aldosterone; (c) norepinephrine; (d) gonadotropins; (e) somatotropin.

14. A cell can increase its own sensitivity to a circulating hormone by the process of: (a) the synergistic effect; (b) the permissive effect; (c) enzyme amplification; (d) the general adaptation syndrome; (e) up-regulation.

15. Which of these is not a paracrine messenger? (a) prostaglandin; (b) lipoxygenase; (c) prostacyclin; (d) leukotriene; (e) thromboxane.

16. All of the following anterior pituitary hormones have a narrow range of targets *except:* (a) GH; (b) FSH; (c) TSH; (d) LH; (e) ACTH.

17. The ___ develop(s) from a pouch in the roof of the embryonic mouth. (a) anterior pituitary; (b) posterior pituitary; (c) thymus; (d) thyroid; (e) parathyroids

18. All of these are caused by hypersecretion of a hormone *except:* (a) acromegaly; (b) diabetes insipidus; (c) Cushing syndrome; (d) androgenital syndrome; (e) Graves disease.

19. ____ from the liver stimulates iron absorption from the small intestine. (a) Hepcidin; (b) Somatomedin; (c) Angiotensinogen; (d) Insulin; (e) Glucagon

20. Which of these is not a correct association between a hormone and its source? (a) leptin-adipose; (b) ANP- heart; (c) IGF-I-liver; (d) ghrelin-stomach; (e) T_3-thymus.

21. Which of these is/are true about hormone actions?
    1. ANP decreases blood pressure by stimulating the kidneys to increase sodium and water excretion.
    2. EPO made in the liver and kidney stimulates white blood cell output during stress.
    3. Angiotensin II is a powerful vasoconstrictor; it also stimulates release of aldosterone.
    4. Aldosterone has a short half-life for a steroid hormone because it binds to transcortin.

    (a) 1 & 3; (b) 2 & 4; (c) 1, 2, & 3; (d) 4 only; (e) all the above

22. Which of these is/are true about the pituitary gland?
    1. The adenohypophysis and neurohypophysis are regions of the anterior pituitary.
    2. The hypophyseal portal system consists of capillaries in the hypothalamus connected to another set of capillaries in the posterior pituitary by venules.
    3. The pars nervosa is connected to the pars distalis by the hypothalamo–hypophyseal tract.
    4. In embryonic development, the hypophyseal pouch grows upward from the pharynx to become the adenohypophysis.

    (a) 1 & 3; (b) 2 & 4; (c) 1, 2, & 3; (d) 4 only; (e) all the above

23. Which of these is/are true?
    1. A hormone can use different second messengers depending on its target cells.
    2. Hormone specificity is partly due to the fact that one hormone can bind only specific receptors on its target cells.
    3. EPI, aldosterone, and A-II increase during an alarm reaction.
    4. NSAIDs can mimic Cushing syndrome.

    (a) 1 & 3; (b) 2 & 4; (c) 1, 2, & 3; (d) 4 only; (e) all the above

24. Which of these is/are true?
    1. Arginine vasopressin (AVP) is a brain neurotransmitter as well as a posterior pituitary hormone.
    2. Insulin release and inhibition are controlled by different neuroendocrine reflex arcs.
    3. Thyroid hormone concentration can inhibit TRH and TSH release.
    4. Hydrophilic hormones are carried in the blood bound to transport proteins.

    (a) 1 & 3; (b) 2 & 4; (c) 1, 2, & 3; (d) 4 only; (e) all the above

25. Which of these is/are true?
    1. Cortisol is an insulin antagonist.
    2. Long-term high cortisol concentrations stimulate the immune system.
    3. A pheochromocytoma mimics the stress response.
    4. The brain produces releasing and inhibiting hormones that target the neurohypophysis.

    (a) 1 & 3; (b) 2 & 4; (c) 1, 2, & 3; (d) 4 only; (e) all the above

E. Word Origins

1. In *adenohypophysis, adeno-* means "adolescent."
2. In *adenohypophysis, -physis* means "growth."
3. In *prolactin, pro-* means "before."
4. In *gonadotropin, trop-* means "absorb."
5. In *oxytocin, toc-* means "secreted."
6. In *calorigenic, calor-* means "heat."
7. In *diabetes mellitus, diabet-* means "sugar."
8. In *polyuria, poly-* means "much, excess."
9. In *glycosuria, -uria* means "urine."
10. In *natriuretic, natri-* means "sodium."
11. -crin
12. -lact
13. diuret-
14. -dipsia
15. melli-

F. Which One Does Not Belong?

1. (a) ACTH → cortisol; (b) TSH → thyroid hormones; (c) GH → somatotropin; (d) LH → estrogen

2. (a) insulin; (b) glucagon; (c) prolactin; (d) thyroid hormones

3. (a) insulin; (b) EPI; (c) cortisol; (d) glucagon

4. (a) diacylglycerol; (b) cAMP; (c) IP$_3$; (d) phosphodiesterase

5. (a) ACTH; (b) LH; (c) PTH; (d) TSH

G. Matching Exercise

Match the disease or condition (1–10) with the causes listed A–P.

A. Hyposecretion of corticosteroids
B. Hyposecretion of ADH
C. Hyposecretion of insulin
D. Hypersecretion of EPI and NE
E. Hypersecretion of thyroid hormones
F. Postpuberty hypersecretion of GH
G. Prepuberty hypersecretion of GH
H. Adult hypothyroidism

I. Lack of dietary iodine
J. Hypersecretion of cortisol
K. Hypersecretion of androgens
L. Hyposecretion of aldosterone
M. Excess ADH
N. Lack of thyroid hormone receptors
O. Hypersecretion of FSH
P. Hypersecretion of insulin

1. Acromegaly
2. Addison disease
3. Cushing syndrome
4. Diabetes insipidus
5. Diabetes mellitus, type 1
6. Endemic goiter
7. Gigantism
8. Myxedema
9. Pheochromocytoma
10. Toxic goiter

18 The Circulatory System: Blood

A. Short Answer

1. Erythrocytes and albumin are mainly responsible for the ___ of blood, its resistance to flow.

2. Production of the formed elements of the blood is called ___, and the tissues that carry this out are called ___ tissues.

3. White blood cells are also called ___.

4. The destiny of the monocytes is to migrate into the tissues and become ___.

5. The nonprotein, iron-containing, oxygen-binding part of the hemoglobin molecule is called the ___.

6. One way of measuring the oxygen-transport capacity of the blood is to centrifuge it and measure the ___, the percentage of the blood volume composed of RBCs.

7. An excessively high RBC count due to cancer of the red bone marrow is called ___.

8. A person whose genotype is $I^A i$ has type ___ blood, and one whose genotype is ii has type ___ blood.

9. Hemolytic disease of the newborn occurs when the mother is blood type ___ and her baby is blood type ___.

10. The most abundant leukocytes in the blood are ___, which function mainly to attack bacteria.

11. A sudden increase in granulocyte count due to cancer of the bone marrow would be classified as ___. [three words]

12. All the mechanisms that limit blood loss from an injured vessel are collectively called ___.

13. The ___ mechanism of blood clotting begins with a clotting factor, ___, released from damaged perivascular tissues.

14. The most abundant clotting protein in the blood is ___, which is converted to the sticky protein ___.

15. The development of blood clots in unbroken blood vessels is called ___. If one of these clots breaks free and begins to travel in the bloodstream, it is called a(n) ___.

B. Matching

A. vascular spasm	G. thrombin	M. megakaryocyte	S. leukopenia
B. leukocytosis	H. coagulation	N. leukemia	T. fibrinogen
C. ferritin	I. thrombocyte	O. hypoxemia	U. hemopoietic stem cell
D. transferrin	J. agglutination	P. granulocyte	V. sickle-cell anemia
E. monocyte	K. erythrocyte	Q. platelet plug	W. transfusion reaction
F. basophils	L. prothrombin	R. factor X	X. blood plasma

1. Quickest mechanism of hemostasis, but not the longest lasting
2. Point at which the intrinsic and extrinsic mechanisms of blood clotting converge
3. Oxygen deficiency in the blood due to various factors
4. A deficiency of white blood cells
5. Source of histamine and heparin
6. Result of an amino acid substitution in hemoglobin
7. Clumping of RBCs by antibodies
8. Protein–iron complex stored in the liver
9. Bone marrow cell that produces platelets
10. Develops from a reticulocyte

C. True or False

1. Most hemophilia diseases are sex-linked recessive conditions in which clotting factors are absent.

2. An Rh+ woman cannot have a baby with erythroblastosis fetalis.

3. Anemia is often caused by a deficiency of oxygen in the blood.

4. Bile pigments come from the breakdown of the globin moiety of hemoglobin.

5. *Hemopoiesis* refers to the production of any of the formed elements—RBCs, WBCs, and platelets.

6. The anti-B antigen is present mainly in individuals with type B blood.

7. The greatest contributor to the blood's colloid osmotic pressure is Na^+.

8. Most of the blood plasma is protein.

9. Basophils are the rarest type of WBCs.

10. All formed elements of the blood arise ultimately from hemopoietic stem cells.

D. Multiple Choice

1. Which of these is a granulocyte? (a) monocyte; (b) lymphocyte; (c) macrophage; (d) eosinophil; (e) erythrocyte

2. Which of these would most likely cause hemolytic anemia? (a) mushroom poisoning; (b) hemophilia; (c) iron deficiency; (d) folic acid deficiency; (e) alcoholism

3. Which of the following is *not* a component of hemostasis? (a) a platelet plug; (b) a vascular spasm; (c) clot retraction; (d) degranulation; (e) agglutination

4. Blood plasma contains all of the following, whereas blood serum contains all of them *except:* (a) immunoglobulins; (b) albumin; (c) fibrinogen; (d) glucose; (e) electrolytes.

5. Blood clotting can be prevented by chemically binding all of the ___, an essential cofactor for several steps in the clotting process. (a) sodium; (b) calcium; (c) magnesium; (d) potassium; (e) phosphorus

6. Albumin functions as a transporter for all *except:* (a) lipids; (b) calcium; (c) oxygen; (d) thyroid hormones; (e) hydrophobic hormones.

7. Fibrinogen is the substrate for the enzyme: (a) factor VIII; (b) prothrombin; (c) thrombin; (d) fibrin; (e) calcium.

8. In court, a man argues that, based on his own blood type, he could not have fathered a certain type O+ baby. As a member of the jury who knows biology, you believe him because the man has ___ blood.
   (a) type B+; (b) type B−; (c) type O+; (d) type A+; (e) type AB−

9. The greatest contributor to the osmotic pressure of the blood is: (a) calcium; (b) hemoglobin; (c) albumin; (d) erythrocytes; (e) sodium.

10. Platelet degranulation releases a vasoconstrictor called: (a) ADP; (b) histamine; (c) serotonin; (d) heparin; (e) epinephrine.

11. Gallstones can cause increased clotting time by: (a) interfering with vitamin K absorption; (b) leading to a vitamin B_{12} deficiency; (c) causing the death of hemopoietic stem cells; (d) triggering the rapid hemolysis of RBCs; (e) producing an immune response against RBCs.

12. An erythrocyte colony-forming unit is committed to becoming an RBC and has receptors for: (a) colony-stimulating factors; (b) erythropoietin; (c) hemoglobin; (d) interleukins; (e) transferrin.

13. Which of the following is a realistic value for the hematocrit of a healthy adult? (a) 12.5 mg/dL; (b) 47%; (c) 70%; (d) 5.0×10^6 RBCs/μL; (e) 33 g/dL

14. Dietary iron cannot be absorbed from the small intestine unless it is bound to: (a) gastroferritin; (b) a heme group; (c) ferritin; (d) transferrin; (e) vitamin K.

15. Which of the following is the *least* likely effect of sickle-cell anemia? (a) cardiac stress; (b) high blood pressure; (c) tissue necrosis; (d) low blood viscosity; (e) hypoxia

16. Iron is transported in the blood plasma bound to: (a) albumin; (b) haptoglobulin; (c) gamma globulin; (d) transferrin; (e) hemoglobin.

17. All the following are the products of heme decomposition *except:* (a) iron; (b) bile pigments; (c) bilirubin; (d) biliverdin; (e) amino acids.

18. Blood vessels are internally coated with ___, which prevents platelet adhesion and unwanted clotting. (a) serotonin; (b) thromboxane A_2; (c) prothrombin; (d) prostacyclin; (e) plasmin

19. Anemia can be caused by: (a) renal failure; (b) edema; (c) hypoxia; (d) hypothyroidism; (e) iron.

20. The leukocytes that react most quickly to bacterial infections are: (a) neutrophils; (b) eosinophils; (c) basophils; (d) monocytes; (e) lymphocytes.

21. Which of these is/are true about blood disorders?
    1. Leukopenia can be caused by mercury poisoning and AIDS.
    2. Hemophilia is caused by sex-linked recessive genes.
    3. Myeloid leukemia results in increased granulocyte production.
    4. Polycythemia can be caused by cancer or dehydration.

    (a) 1 & 3; (b) 2 & 4; (c) 1, 2, & 3; (d) 4 only; (e) all the above

22. Which of these could cause an increase in EPO release from the liver and kidneys?
    1. Snow skiing at Vail, Colorado
    2. Training for a marathon
    3. An auto accident resulting in severe hemorrhage
    4. Emphysema

    (a) 1 & 3; (b) 2 & 4; (c) 1, 2, & 3; (d) 4 only; (e) all the above

23. Which of these is/are true?
    1. Gastroferritin, transferrin, and ferritin are blood proteins that bind iron absorbed from the small intestine.
    2. Lymphocytes are produced in the spleen, lymph nodes, and bone marrow throughout life.
    3. Lymphocytes differentiate into macrophages.
    4. Fetal hemoglobin binds oxygen better than does adult hemoglobin.

    (a) 1 & 3; (b) 2 & 4; (c) 1, 2, & 3; (d) 4 only; (e) all the above

24. Which of these is/are true about leukocytes?
    1. Monocytes kill parasites by enzymatic digestion.
    2. Lymphocytes secrete antibodies.
    3. Neutrophils release vasodilators.
    4. Eosinophil numbers fluctuate with the phase of a woman's menstrual cycle.

    (a) 1 & 3; (b) 2 & 4; (c) 1, 2, & 3; (d) 4 only; (e) all the above

25. Which of these is/are true?
    1. An allergic response to ragweed pollen triggers leukopoiesis, increasing eosinophil number.
    2. Leeches are still used today after certain types of surgery.
    3. People who have sickle-cell trait are resistant to malaria.
    4. Giving the wrong blood type during a transfusion causes immediate death.

    (a) 1 & 3; (b) 2 & 4; (c) 1, 2, & 3; (d) 4 only; (e) all the above

E. Word Origins

1. In *hematology, hemat-* means "blood."
2. In *erythrocyte, erythro-* means "blood."
3. In *hemopoiesis, -poiesis* means "bone marrow."
4. In *myelocytic, myelo-* means "bone marrow."
5. In *hypoxemia, -emia* means "blood condition."
6. In *hemolysis, -lysis* means "splitting, breaking up."
7. In *anemia, an-* means "without."
8. In *leukocytosis, -osis* means "insufficient."
9. In *hemostasis, -stasis* means "stage."
10. In *hematoma, -oma* means "coagulation."
11. leuko-
12. verd-
13. rub-
14. thrombo-
15. -crit

F. Which One Does Not Belong?

1. (a) eosinophil; (b) monocyte; (c) basophil; (d) platelet

2. (a) copper; (b) vitamin B_{12}; (c) oxygen; (d) iron

3. (a) platelet plug formation; (b) thrombus; (c) coagulation; (d) vascular spasm

4. (a) ammonia; (b) uric acid; (c) urea; (d) amino acid

5. (a) megakaryocyte → T lymphocyte; (b) myeloblast → eosinophil; (c) reticulocyte → erythrocyte; (d) myeloblast → basophil

19   The Circulatory System: The Heart

A.   Short Answer

1. The circulatory system has two divisions: a(n) ___ circuit that furnishes blood to the entire body and a(n) ___ circuit that functions only for gas exchange with the air in the lungs.

2. The heart is enclosed in a fibrous, two-layered sac called the ___.

3. Obstruction of a coronary artery leads to ___, sudden death of a part of the heart muscle.

4. The ventricles have ___ muscles that anchor stringy ___ to prevent the atrioventricular valves from prolapsing during ventricular systole.

5. The ___ and ___ veins lie in interventricular sulci alongside the interventricular arteries.

6. ___ occurs when different areas of the heart muscle receive electrical stimuli at different times.

7. Compared to skeletal muscle, cardiac muscle has a relatively scanty ___ and, therefore, gets most of its calcium ions from the ___ during contraction.

8. Unlike skeletal muscle fibers, cardiocytes are joined end-to-end by mechanical and electrical junctions called ___.

9. Normally, each cycle of cardiac contraction is set off by its pacemaker, the ___.

10. When the left heart fails in congestive heart failure, ___ edema occurs.

11. Unlike those of skeletal muscle and neurons, the action potentials of cardiac muscle exhibit a plateau resulting from a prolonged influx of ___ ions.

12. The ventricles begin to relax and the second heart sound is heard at the time of the ___ wave, or complex, of the electrocardiogram.

13. The contraction of any heart chamber is called ___, and its relaxation is called ___.

14. The amount of blood ejected by one contraction of a ventricle is called its ___, and the percentage of its blood ejected is called its ___.

15. Agents such as epinephrine are said to have a positive ___ effect because they increase the heart rate and a positive ___ effect because they increase the force of its contractions.

B. Matching

A. afterload   G. calcium influx   M. QRS complex   S. tricuspid valve
B. desmosome   H. arrhythmia   N. residual volume   T. T wave
C. P wave   I. stroke volume   O. AV bundle   U. sinus rhythm
D. sodium influx   J. chronotropic   P. gap junctions   V. end-systolic volume
E. ectopic focus   K. mitral valve   Q. preload   W. end-diastolic volume
F. nodal rhythm   L. inotropic   R. prolapse   X. semilunar valve

1. All the forces that oppose ejection of blood from the ventricles
2. The left atrioventricular valve
3. Abnormal heartbeat
4. The intercellular electrical connections at an intercalated disc
5. May take over for a damaged SA node
6. End-diastolic volume minus stroke volume
7. The normal heartbeat governed by the SA node
8. Opens during ventricular systole
9. Occurs at the time of the first heart sound
10. Amount of blood in a ventricle just before the onset of contraction

C. True or False

1. An increased afterload increases the heart's ejection fraction.

2. Anything that exerts a positive inotropic effect on the heart is likely to increase its ejection fraction.

3. Cells of the SA and AV nodes are the only cardiac cells that depolarize spontaneously and rhythmically.

4. Cardiac nerves carry sympathetic, parasympathetic, and sensory fibers.

5. The AV node and Purkinje fibers conduct signals faster than any other cardiac tissue.

6. Because athletes are in better-than-average physical condition, their hearts beat faster at rest.

7. The coronary sinus empties into the right atrium.

8. The left ventricle, being larger than the right, has a higher stroke volume.

9. The semilunar valves prevent ventricular blood from regurgitating back into the atria.

10. An atheroma begins as a complex of lipids, smooth muscle, and macrophages that, when calcified, lead to arteriosclerosis.

D. Multiple Choice

1. Causes of CHF include all *except*: (a) valvular defects; (b) ascites; (c) chronic hypertension; (d) myocardial infarction; (e) congenital heart defects.

2. Ineffective tendinous cords would most likely cause: (a) valvular stenosis; (b) fibrillation; (c) a heart block; (d) valvular prolapse; (e) a silent myocardial infarction.

3. Blood normally leaves the right ventricle by way of the: (a) pulmonary semilunar valve; (b) right AV valve; (c) tricuspid valve; (d) bicuspid valve; (e) aortic semilunar valve.

4. A heart rate of 45 bpm and an absence of P waves from the ECG suggests: (a) ventricular fibrillation; (b) arrhythmia; (c) heart block; (d) damage to the SA node; (e) cardiac auscultation.

5. The rising phase of the action potential in SA node cells results from: (a) the opening of slow calcium channels; (b) the closing of potassium channels; (c) calcium influx; (d) potassium influx; (e) sodium efflux.

6. The atria contract during the: (a) first heart sound; (b) second heart sound; (c) QRS complex; (d) P-Q segment; (e) S-T segment.

7. When you are resting, atrial systole adds about ___ of the ESV to the ventricles. (a) 5%; (b) 12%; (c) 30%; (d) 50%; (e) 75–80%

8. In the action potential of a cardiocyte, calcium gates close: (a) when the pacemaker potential reaches threshold; (b) when the voltage reaches its highest positive value; (c) just as the cell begins to depolarize; (d) at the end of the plateau; (e) at the end of repolarization.

9. In the heart, acetylcholine has a: (a) positive inotropic effect; (b) negative chemotropic effect; (c) cardiotonic effect; (d) negative inotropic effect; (e) negative chronotropic effect.

10. Without vagal stimulation of the SA node, the heart beats at an intrinsic rate of ___ beats per minute. (a) 100; (b) 20; (c) 70; (d) 80; (e) 200

11. ___ occurs in ventricular systole before the semilunar valves open. (a) Isovolumetric relaxation; (b) The T wave; (c) The second heart sound; (d) Isovolumetric contraction; (e) Rapid ventricular filling

12. Which of the following is *not* an anatomical feature of the heart? (a) thebesian veins; (b) aortic semilunar valve; (c) trabeculae carneae; (d) pulmonary tricuspid valve; (e) papillary muscles

13. How many pulmonary veins open into the right atrium? (a) 0; (b) 1; (c) 2; (d) 4; (e) 8

14. If a person's EDV is 140 mL, the ESV is 55 mL, and the ejection fraction is 0.61, then the stroke volume must be about: (a) 55 mL; (b) 85 mL; (c) 61 mL; (d) 34 mL; (e) 195 mL.

15. Given the stroke volume calculated in question 14 and a cardiac output of 6.6 L/min, this person must have a heart rate of about: (a) 47 bpm; (b) 120 bpm; (c) 70 bpm; (d) 77 bpm; (e) it cannot be calculated from the information given.

16. Intercalated discs are characterized by all *except:* (a) desmosomes; (b) mitochondria; (c) fascia adherens; (d) interdigitating folds; (e) gap junctions.

17. Myocardial infarctions might be much more common if it were not for: (a) the heart's double circulation; (b) the tendinous cords; (c) arterial anastomoses; (d) the trabeculae carneae; (e) the intercalated discs.

18. An enlarged Q wave on an EKG indicates: (a) premature ventricular contraction; (b) ventricular hypertrophy; (c) heart block; (d) myocardial infarction; (e) atrial hypertrophy.

19. Nicotine causes an increase in heart rate by: (a) stimulating epinephrine and norepinephrine release; (b) increasing acetylcholine release; (c) stimulating hypercapnia; (d) decreasing preload; (e) breaking the Frank-Starling law of the heart.

20. The dicrotic notch in the aortic pressure curve represents: (a) a pressure jump caused by the opening of the AV valves; (b) a drop in pressure caused by the opening of the mitral valve; (c) the diastolic blood pressure; (d) blood regurgitation into the left ventricle until the semilunar valve closes; (e) the systolic blood pressure.

21. Which of these is/are true about control over the heart rate?
    1. Acetylcholine hyperpolarizes pacemaker cells by opening K^+ channels.
    2. When blood pressure decreases, the cardioacceleratory center in the brain stimulates an increase in heart rate.
    3. Acidosis and hypercapnia both increase heart rate.
    4. Stimulants in chocolate increase heart rate by inhibiting cAMP breakdown.

    (a) 1 & 3; (b) 2 & 4; (c) 1, 2, & 3; (d) 4 only; (e) all the above

22. Which of these is/are true about the general features of the heart?
    1. The heart lies along the median plane, with the aortic arch leaving the right side of the mediastinum.
    2. Blood enters the heart on the right side and leaves on the left.
    3. The heart has a fibrous skeleton, so called because of its framework of cartilaginous fibers.
    4. The four valves in the heart allow one-way blood flow.

    (a) 1 & 3; (b) 2 & 4; (c) 1, 2, & 3; (d) 4 only; (e) all the above

23. Which of these is/are true about electrical activity in the heart?
    1. Action potentials in cardiocytes are very similar to those in pacemaker cells.
    2. Hyperkalemia slows the heart rate.
    3. The vagus nerve releases potassium that slows the heart rate.
    4. An action potential in a cardiocyte lasts much longer than one in a skeletal muscle cell.

    (a) 1 & 3; (b) 2 & 4; (c) 1, 2, & 3; (d) 4 only; (e) all the above

24. Which of these is/are true about fluid dynamics in the heart?
    1. Increasing pressure in a heart chamber decreases its volume.
    2. Most ventricular filling occurs during ventricular diastole.
    3. Ventricular ejection follows isovolumetric contraction.
    4. Atrial filling increases with increased ventricular filling.

    (a) 1 & 3; (b) 2 & 4; (c) 1, 2, & 3; (d) 4 only; (e) all the above

25. Which of these is/are true?
    1. *Atherosclerosis* and *arteriosclerosis* are two terms used for the same disease.
    2. About half the deaths in the United States are caused by myocardial infarctions.
    3. The mammalian heart cannot beat if all nerves to it are cut.
    4 Cardiac muscle can use amino acids, glucose, fatty acids, and ketones as fuel for contraction.

    (a) 1 & 3; (b) 2 & 4; (c) 1, 2, & 3; (d) 4 only; (e) all the above

E. Word Origins

1. In *cardiology, cardi-* means "heart."
2. In *pericardium, peri-* means "throughout."
3. In *ventricle, ventr-* means "chamber."
4. In *semilunar, semi-* means "half."
5. In *coronary, -ary* means "unlike."
6. In *infarction, infarct-* means "dead."
7. In *syncytium, cyt-* means "cell."
8. In *autorhythmic, auto-* means "self."
9. In *ectopic, ec-* means "out of."
10. In *tachycardia, tachy-* means "too slow."
11. steno-
12. trabecula-
13. sphygmo-
14. angina
15. -sclerosis

F. Which One Does Not Belong?

1. (a) sedentary lifestyle; (b) smoking; (c) gender; (d) stress

2. (a) circumflex branch of the LCA; (b) right coronary artery; (c) anterior interventricular branch of the LCA; (d) coronary sinus

3. (a) tendinous cords; (b) Purkinje fibers; (c) SA node; (d) AV node

4. (a) preload; (b) heart rate; (c) contractility; (d) afterload

5. (a) S-T segment, ventricular systole; (b) P wave, atrial depolarization; (c) QRS interval, atrial repolarization; (d) T wave, ventricular contraction

G. Figure Exercises

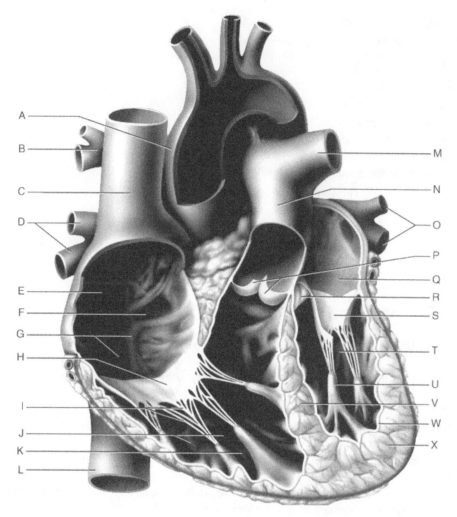

I. Match the structure in the heart with the following descriptions. Some questions may have more than one answer and some letters may be used more than once.

1. Vessel(s) that carry/carries blood with high oxygen levels
2. Open(s) when ventricular pressure is high
3. Become(s) taut and stabilize(s) valve flaps during systole
4. Contain(s) the SA node
5. Receive(s) blood from lungs
6. Remnant of embryonic shunt that bypasses lungs between atria
7. Location of bundle branches
8. Carry/carries blood to all organs except lungs
9. Drain(s) body regions above heart
10. Pulmonary pump

II.   Unscramble the following structures to indicate the correct pathway through the heart. Begin
      with the right atrium.

A. Aorta	I. Aortic semilunar valve
B. Inferior and superior venae cavae	J. Left atrium
C. Left ventricle	K. Lungs
D. Mitral valve	L. Pulmonary arteries
E. Pulmonary semilunar valve	M. Pulmonary trunk
F. Right atrium	N. Pulmonary veins
G. Systemic arteries	O. Right ventricle
H. Tricuspid valve	P. Systemic veins

20   The Circulatory System: Blood Vessels and Circulation

A. Short Answer

1. Blood leaving the heart usually flows through arteries, a capillary bed, veins, and back to the heart but, in a(n) ___, it flows through two capillary beds in succession before returning to the heart.

2. Arteries and veins have three-layered walls. The innermost layer is called the ___, and it is lined by a simple squamous epithelium called the ___.

3. With minor exceptions, ___ are the only blood vessels that can exchange substances with the surrounding tissues.

4. The ___ reflex is an ANS response to a decrease in blood flow to the brain.

5. In a blood pressure such as 120/75, 120 represents the ___ pressure and 75 represents the ___ pressure.

6. The subclavians and common iliacs are examples of ___ arteries, while the femorals and renals are examples of ___ arteries.

7. Blood flow is a function of pressure divided by ___. The most effective way of changing that variable is to change the ___ of the blood vessels.

8. ___ is the ability of a tissue to control its own blood flow, as opposed to control by hormonal and neural mechanisms.

9. Capillary filtration is opposed mainly by ___ pressure—the difference between the colloid osmotic pressure of the blood and that of the tissue fluid.

10. In ___, the heart does not pump enough blood to meet the body's metabolic needs, not because of cardiac failure or a loss of blood from the body but because the blood has accumulated in the extremities.

11. The inferior end of the aorta branches into right and left ___.

12. The third and last artery to branch from the aortic arch is the ___ artery.

13. The ___ artery arises from the abdominal aorta and supplies the small intestine with most of its blood.

14. The subclavian vein inferior to the clavicle and the internal jugular vein of the neck join each other and form the ___ vein.

15. The ___ vein enters the hilum of the liver, carrying most of the nutrients absorbed by the small intestine.

B. Matching

A. vasomotion	G. filtration	M. transcytosis	S. brachiocephalic v.
B. fenestration	H. portal system	N. vasa vasorum	T. precapillary sphincter
C. ischemia	I. peripheral resistance	O. varicose vein	U. MAP
D. pulse pressure	J. brachiocephalic a.	P. afterload	V. inotropic effect
E. angiogenesis	K. baroreflex	Q. aneurysm	W. external jugular v.
F. veins	L. sinus	R. axillary a.	X. reactive hyperemia

1. Increase in blood flow after blood supply has been cut off
2. Regulates the admission of blood into a capillary
3. The only blood vessels with valves
4. Opposition to blood flow away from the heart
5. Diastolic BP + (systolic – diastolic BP)/3
6. Means of moving fluid from one side of the capillary endothelial cell to the other
7. Most effective way of producing short-term changes in blood flow
8. Initiated by stretch receptors in the heart wall and major arteries
9. Connection from the subclavian to the brachial artery
10. Insufficient blood supply

C. True or False

1. Blood doesn't always go through capillaries on the way from an artery to a vein.

2. An aneurysm is a ruptured blood vessel.

3. Anaphylactic shock is a form of venous pooling shock.

4. Angiotensin II is a powerful vasodilator.

5. Aldosterone helps support blood pressure by promoting sodium retention.

6. Blood capillaries must reabsorb all the fluid they emit, or else edema will occur.

7. Hypoproteinemia will usually cause edema.

8. The one-way flow of blood through the arteries is ensured by a series of arterial valves.

9. If the radius of a blood vessel doubles and all other factors remain the same, blood flow through that vessel also doubles.

10. Blood reaches the crural region only by way of the external iliac artery.

D. Multiple Choice

1. Blood normally enters a capillary bed from a: (a) vein; (b) distributing artery; (c) conducting artery; (d) metarteriole; (e) thoroughfare channel.

2. Where a high rate of filtration is important, as in the kidneys, we are likely to see: (a) continuous capillaries; (b) discontinuous capillaries; (c) arteriovenous capillaries; (d) fenestrated capillaries; (e) porous capillaries.

3. Arteriovenous ___ are routes by which some blood bypasses the capillaries. (a) fenestrations; (b) reservoirs; (c) anastomoses; (d) occlusions; (e) sinuses

4. The growth of new blood vessels into an ischemic tissue is called: (a) angiogenesis; (b) reactive hyperemia; (c) adaptive hyperemia; (d) the ischemic reflex; (e) the Frank-Starling effect.

5. The secretion of ___ causes the blood pressure to drop. (a) serotonin; (b) epinephrine; (c) norepinephrine; (d) angiotensin II; (e) histamine

6. Blood flow will decrease if: (a) vessel radius increases; (b) viscosity increases; (c) the value of π increases; (d) pressure increases; (e) afterload decreases.

7. Pulse pressure is high in the aorta and lower in the arterioles mainly because of: (a) a difference in cardiac output; (b) the sheer number of arterioles; (c) the distensibility and elasticity of the arteries; (d) the smaller diameter of the arterioles; (e) the effect of the muscular pump.

8. The vasa vasorum are: (a) fenestrated capillaries; (b) small vessels that supply blood to the tissues of large vessels; (c) arteriovenous anastomoses; (d) portal systems; (e) the three layers of a blood vessel.

9. All of these are true about carotid sinuses *except:* (a) they respond to an increase in blood pressure; (b) they are baroreceptors; (c) they are innervated by the ninth cranial nerve; (d) they are sensitive to pH; (e) they are stretch receptors.

10. Excessive fluid loss by perspiration or hemorrhage is most likely to cause: (a) anaphylactic shock; (b) cardiac shock; (c) neurogenic shock; (d) obstructed venous return shock; (e) hypovolemic shock.

11. All of the following are branches of the external carotid artery *except* the: (a) basilar artery; (b) superior thyroid artery; (c) occipital artery; (d) lingual artery; (e) facial artery.

12. There are circumflex arteries in the coronary circulation and: (a) surrounding the pituitary gland; (b) in the hepatic portal system; (c) on the lesser curvature of the stomach; (d) at the base of the hand; (e) surrounding the upper humerus.

13. We have two of each of the following vessels *except:* (a) facial vein(s); (b) hepatic vein(s); (c) common iliac vein(s); (d) brachiocephalic artery/arteries; (e) brachiocephalic vein(s).

14. The brachial artery ends near the elbow, where it gives rise to the: (a) radial and ulnar arteries; (b) deep brachial and radial recurrent arteries; (c) radial and radial recurrent arteries; (d) radial and anterior interosseous arteries; (e) ulnar and ulnar recurrent arteries.

15. All are branches of the aorta *except* the: (a) azygos artery; (b) left common carotid artery; (c) celiac trunk; (d) superior mesenteric arteries; (e) common iliac arteries.

16. All are true of hypertension *except:* (a) it causes stroke and kidney failure; (b) it leads to atherosclerosis; (c) it is caused by a high-salt diet; (d) nicotine compounds its effects; (e) uncontrollable risk factors include race, gender, and age.

17. Stimulation of the aortic and carotid bodies results in all *except*: (a) increased systemic blood pressure; (b) increased hypercapnia; (c) increased lung perfusion; (d) increased breathing rate; (e) increased vasomotion.

18. A person with a blood pressure of 145/95 has a mean arterial blood pressure of about ___ and is at risk for kidney failure and atherosclerosis. (a) 230 mmHg; (b) 100 mmHg; (c) 117 mmHg; (d) 50 mmHg; (e) we cannot tell without knowing the heart rate.

19. When we say the bloodstream shows laminar flow, we mean: (a) the blood near the center of the vessel lumen flows faster than the blood near the walls; (b) arterial blood flows in a pulsatile fashion, faster during systole and slower during diastole; (c) venous blood flows at a constant speed; (d) blood flow is smooth and silent; (e) blood flow is aided by lamellae of the venous valves.

20. The baroreflex causes: (a) a decrease in sympathetic tone when blood pressure rises; (b) an increase in sympathetic tone when blood pressure rises; (c) vasoconstriction in response to a rise in blood pressure; (d) a decrease in cardiac output when blood pressure drops; (e) a decrease in heart rate when blood pressure drops.

21. Which of these is/are true about capillary dynamics?
    1. Congestive heart failure can cause edema.
    2. Filtration and reabsorption rates must be equal.
    3. Colloid osmotic pressure is a measure of proteins in the plasma, and it is important in capillary function.
    4. Edema is most often caused by decreased venous pressure.

    (a) 1 & 3; (b) 2 & 4; (c) 1, 2, & 3; (d) 4 only; (e) all the above

22. Which of these is/are true about shock?
    1. The baroreflex counteracts shock.
    2. Fainting is a way of overcoming shock.
    3. In vascular shock, blood volume is normal, but blood pools in the limbs.
    4. Myocardial infarctions can lead to low-venous-return shock.

    (a) 1 & 3; (b) 2 & 4; (c) 1, 2, & 3; (d) 4 only; (e) all the above

23. Which of these is/are true?
    1. Conducting and elastic arteries respond to pumping of the ventricles, expanding when the ventricles contract.
    2. Blood proteins synthesized in the liver enter the blood through sinusoids and add to colloid osmotic pressure.
    3. Hemorrhoids are varicose veins.
    4. Capillary anatomy is well suited to the capillaries' function of linking arteries to veins.

    (a) 1 & 3; (b) 2 & 4; (c) 1, 2, & 3; (d) 4 only; (e) all the above

24. Which of these is/are true about venous pressure and venous return?
    1. The most important factor in venous blood flow regulation is the pressure produced by the heart's pumping action.
    2. The skeletal muscle and respiratory pumps aid in venous return.
    3. Venous return increases with exercise.
    4. A low venous return rate can lead to shock.

    (a) 1 & 3; (b) 2 & 4; (c) 1, 2, & 3; (d) 4 only; (e) all the above

25. Which of these is/are true?
    1. The force blood places on a vessel wall is called perfusion.
    2. Arteries have pulsatile blood flow, while veins do not.
    3. The greater the vessel radius at any given time, the greater is its resistance.
    4. Blood pressure is determined by blood volume, cardiac output, and resistance.

    (a) 1 & 3; (b) 2 & 4; (c) 1, 2, & 3; (d) 4 only; (e) all the above

E. Word Origins

1. In *adventitia, advent-* means "arrival."
2. In *vasa vasorum, vas-* means "vein."
3. In *fenestrated, fenestr-* means "window."
4. In *metarteriole, met-* means "to join."
5. In *angiogenesis, angio-* means "vessel."
6. In *baroreflex, baro-* means "air."
7. In *brachiocephalic, brachio-* means "arm."
8. In *thyrocervical, cervi-* means "shoulder."
9. In *carotid, carot-* means "conscious."
10. In *phrenic, phren-* means "mind."
11. lamina-
12. acr-
13. costo-
14. zygo-
15. celi-

F. Which One Does Not Belong?

1. (a) superior mesenteric artery; (b) celiac trunk; (c) external iliac artery; (d) aorta

2. (a) filtration; (b) diffusion; (c) transcytosis; (d) active transport

3. (a) pulsatile flow; (b) vessel length; (c) blood viscosity; (d) vessel radius

4. (a) skeletal muscle pump; (b) gravity; (c) respiratory pump; (d) capillary filtration

5. (a) epinephrine; (b) atrial natriuretic peptide; (c) antidiuretic hormone; (d) angiotensin II

G. Blood Vessel Exercises

I. For each vessel below name the vessel(s) into which blood will flow next in the circulatory pathway.
    Example: right common carotid a. → right internal and external carotid aa.

   1.   right subclavian a. → _____

   2.   right popliteal vein → _____

   3.   azygos vein → _____

   4.   left vertebral a. → _____

   5.   left common iliac a. → _____

   6.   right great saphenous vein → _____

   7.   brachiocephalic trunk → _____

   8.   brachiocephalic vein → _____

   9.   celiac trunk → _____

  10.   hepatic portal vein → _____

II. List 10 branches of the aorta.

   1.   _____

   2.   _____

   3.   _____

   4.   _____

   5.   _____

   6.   _____

   7.   _____

   8.   _____

   9.   _____

  10.   _____

21 The Lymphatic and Immune Systems

A. Short Answer

1. The lymphatic system collects excess tissue fluid and returns it to the bloodstream at the ___ *subclavian* veins.

2. ___ are microbes like bacteria, viruses, and fungi that cause disease.

3. The lymphatic organs best positioned to detect inhaled and ingested pathogens are ___.

4. Neutrophils undergo a reaction, called the ___, to produce strong oxidizing agents that kill nearby bacteria.

5. Redness, heat, and pain are some of the classic signs of a defensive reaction called ___.

6. Mast cells and basophils secrete an anticoagulant called ___ and a vasodilator called ___.

7. Cells infected with viruses may secrete chemical messages, called ___, which activate natural killer cells that interfere with viral invasion of healthy cells.

8. Pathogens in the lymph are often detected as it passes through organs called the ___.

9. The hormone ___ suppresses the immune system partly because it inhibits T cell and macrophage activity.

10. Helper T cells respond only to antigen fragments that are displayed by various ___ cells.

11. In ___, C3b makes pathogens more "appetizing" to phagocytes.

12. ___ is a defense mechanism that does not require prior exposure to a pathogen or disease agent.

13. ___ is a process in which antibody molecules cause clumping of foreign cells or antigen molecules.

14. The only class of T lymphocytes that directly attack foreign cells is the ___ cells.

15. ___ is a severe circulatory failure that results when an antigen is introduced into the bloodstream of a hypersensitive person.

B. Matching

A. diapedesis
B. basophil
C. neutralization
D. regulatory T cells
E. MHC protein
F. perforin

G. plasma cell
H. lymph node
I. clonal selection
J. interleukin
K. interferon
L. hyaluronidase

M. histamine
N. hapten
O. spleen
P. helper factor
Q. bradykinin
R. margination

S. thymus gland
T. eosinophil
U. immunocompetence
V. clonal deletion
W. T cell recall response
X. cytotoxic T cells

1. Lymphocyte source and RBC reservoir
2. Limit immune response by inhibiting multiplication of T cells
3. Protein made by NK and T_c cells that makes a hole in target cells leading to their apoptosis
4. Results in production of effector cells and memory T cells
5. Migration of WBCs through the capillary endothelium
6. Stimulates pain receptors in inflamed tissue
7. Site where T lymphocytes become immunocompetent
8. Lymphatic organ with both afferent and efferent lymphatic vessels
9. Secreted by some pathogenic organisms to liquefy the tissue gel
10. Phagocytizes antigen–antibody complexes

C. True or False

1. Antigens are usually large molecules with highly repetitive structures.

2. Clonal deletion destroys B and T cells that would otherwise react against the body's own antigens.

3. Helper T cells cannot recognize free antigen molecules, only antigens that are bound to MHC proteins.

4. Granzymes produced by NK cells cause apoptosis in enemy cells.

5. Macrophages not only phagocytize tissue debris and foreign cells but also present antigens, so that helper T cells can recognize them.

6. Opsonization is a process in which foreign cells are lysed by complement proteins.

7. Plasma cells develop from B cells but are larger and have more rough endoplasmic reticulum.

8. Plasma cells synthesize the blood plasma.

9. Anaphylactic shock is characterized by dangerously high blood pressure.

10. All the antibody molecules in one person have identical heavy chains but different light chains.

D. Multiple Choice

1. Splenic tissue called ___ forms sleeves containing lymphocytes and macrophages around branches of the splenic artery. (a) follicles; (b) crypts; (c) trabeculae; (d) red pulp; (e) white pulp

2. All of the following are examples of nonspecific resistance *except:* (a) the stratum corneum; (b) lysozyme; (c) the anamnestic response; (d) fever; (e) immune surveillance.

3. Plasma cells that secrete antibodies arise from: (a) B lymphocytes; (b) killer T cells; (c) helper T cells; (d) mast cells; (e) macrophages.

4. The antigen-binding sites of an antibody molecule are found: (a) at the tips of the constant regions; (b) at the tips of the variable regions; (c) on the light chains only; (d) on the heavy chains only; (e) on the antigenic determinants.

5. All of the following are involved in specific immunity *except* ___, which function only in nonspecific defense. (a) B cells; (b) killer T cells; (c) helper T cells; (d) antigen-presenting cells; (e) natural killer cells

6. All of the following are lymphatic organs *except:* (a) red bone marrow; (b) the liver; (c) the spleen; (d) the thymus; (e) the tonsils.

7. Circulating antibodies are of central importance in: (a) cell-mediated immunity; (b) passive immunity; (c) nonspecific defense; (d) humoral immunity; (e) diplomatic immunity.

8. ___ occurs when metastasizing cancer cells become trapped in lymph nodes. (a) An allergy; (b) Lymphadenitis; (c) A lymphoma; (d) Lupus erythematosus; (e) Hodgkin disease

9. Which of the following are not APC's? (a) dendritic cells; (b) memory T cells; (c) macrophages; (d) B cells; (e) reticular cells

10. Most circulating antibodies, and the only ones that easily cross the placenta, are in the class: (a) IgA; (b) IgM; (c) IgG; (d) IgD; (e) IgE.   IgG

11. Lymphocytes are said to be ___ once they have developed their surface receptors for an antigen. (a) capped; (b) opsonized; (c) cloned; (d) sensitized; (e) immunocompetent

12. Tapeworms, hookworms, and other parasites too large to be phagocytized are attacked mainly by: (a) eosinophils; (b) NK cells; (c) cytotoxic T cells; (d) macrophages; (e) neutrophils.

13. The respiratory burst of a neutrophil produces ___, which is highly toxic to nearby bacteria. (a) perforin; (b) hypochlorite; (c) complement; (d) interleukin; (e) interferon

14. Which of the following are most important as antigen-presenting cells? (a) NK cells; (b) T cells; (c) cytotoxic T cells; (d) eosinophils; (e) macrophages

15. HIV attacks and destroys ___, the central coordinating cells of humoral immunity, cellular immunity, and some forms of nonspecific defense. (a) B lymphocytes; (b) plasma cells; (c) antigen-presenting cells; (d) helper T cells; (e) myelocytic stem cells

16. Inhaled pathogens are intercepted by the: (a) thymus; (b) thyroid; (c) spleen; (d) pharyngeal tonsil; (e) cisterna chyli.

17. Artificial active immunity results from: (a) exposure to antigens in the environment; (b) histocompatibility; (c) antigenic determinants; (d) vaccination against a disease; (e) positive selection.

18. ____ wander through connective tissues and kill bacteria by phagocytosis or the respiratory burst. (a) Lymphocytes; (b) Macrophages; (c) Basophils; (d) Eosinophils; (e) Neutrophils

19. External barriers in nonspecific resistance include all *except:* (a) lysozyme; (b) acid mantle; (c) phagocytes; (d) defensins; (e) mucus.

20. Somatic recombination is a process responsible for: (a) immunological self-tolerance; (b) the T cell recall response; (c) antibody diversity; (d) clonal selection; (e) complement reactions.

21. Which of these is/are true about AIDS?
    1. The virus that causes it attacks helper T cells as well as dendritic cells and macrophages.
    2. It can be transmitted from mother to fetus via the placenta.
    3. It is caused by a retrovirus.
    4. Most AIDS patients contract the disease through homosexual contact.

    (a) 1 & 3; (b) 2 & 4; (c) 1, 2, & 3; (d) 4 only; (e) all the above

22. The functions of the lymphatic system include:
    1. absorption of lipids from digestion.
    2. immunity.
    3. absorption of fluid and proteins lost by blood capillaries.
    4. production of lymphocytes.

    (a) 1 & 3; (b) 2 & 4; (c) 1, 2, & 3; (d) 4 only; (e) all the above

Preload – diastolic
Afterload – systolic

23. Which of these is/are true?
    1. Plasma cells are derived from T cells.
    2. Antibodies destroy antigens.
    3. IgE attracts basophils to sites of parasitic infection.
    4. Asthma is more common in children living in ultraclean environments than in those living on farms.

    (a) 1 & 3; (b) 2 & 4; (c) 1, 2, & 3; (d) 4 only; (e) all the above

24. Fever:
    1. is an adaptive defense mechanism.
    2. is often caused by exogenous pyrogens.
    3. involves prostaglandins that reset the body temperature set point.
    4. is normally beneficial.

    (a) 1 & 3; (b) 2 & 4; (c) 1, 2, & 3; (d) 4 only; (e) all the above

25. Which of these is/are true?
    1. Epitopes stimulate immune responses. √
    2. MHC proteins are unique to each individual. √
    3. An example of a hapten is penicillin. √
    4. Lymph capillaries anastomose with blood capillaries in order to prevent edema.

    (a) 1 & 3; (b) 2 & 4; (c) 1, 2, & 3; (d) 4 only; (e) all the above

E. Word Origins

1. In *lymphadenopathy, -pathy* means "swollen."
2. In *immunology, immuno-* means "infection."
3. In *pyrogen, pyro-* means "fire, heat."
4. In *diapedesis, dia-* means "through."
5. In *antibody, anti-* means "against."
6. In *interleukin, inter-* means "within."
7. In *allergy, -allo* means "next to."
8. In *anamnestic, ana-* means "once."
9. In *anamnestic, mnes-* means "remember."
10. In *anaphylaxis, phylax-* means "infection."
11. -gen
12. -lysis
13. patho-
14. opson-
15. -itis

F. Which One Does Not Belong?

1. (a) produces fetal blood cells; (b) disposes of erythrocyte; (c) makes plasma cells; (d) rids blood of bacteria

2. (a) natural killer cells; (b) microglia of the CNS; (c) dendritic cells in oral mucosa; (d) histiocytes of loose connective tissue

3. (a) pathogen containment; (b) antigen presentation; (c) tissue cleanup; (d) mobilization of defenses

4. (a) memory cell; (b) cytotoxic T cell; (c) plasma cell; (d) helper cells

5. (a) agglutination; (b) diapedesis; (c) neutralization; (d) precipitation

22   The Respiratory System

A. Short Answer

1. The ___ division of the respiratory system exchanges gases with the blood, while the ___ division only carries air to and from the blood.

2. The nasal ___ divides the nasal cavity into right and left chambers, called the nasal ___.

3. The ___ and the ___ of the larynx prevent food and drink from entering the airway.

4. Each lobe of the lung receives air from one ___; thus, there are three of these for the right lung and two of them for the left.

5. Resistance to airflow is adjusted mainly at the level of the ___, which have smooth muscle but no cartilage in their walls.

6. Gas exchange occurs across the walls of minute air sacs called ___.

7. The primary respiratory rhythm generator is the ___.

8. Inspiration occurs when the pressure between the parietal and visceral ___ is lower than the atmospheric pressure outside the body.

9. Great alveolar (type II) cells of the alveoli secrete a(n) ___ that decreases surface tension and prevents alveolar collapse when exhaling.

10. Vital capacity can be measured with a device called a(n) ___.

11. Difficulty breathing is called ___, and the heavy breathing that occurs during exercise is called ___.

12. Boyle law describes the relationship between pressure and volume of gases.

13. Henry law describes the factors that determine how much gas will dissolve in a liquid, such as the blood plasma.

14. ___ coupling is the adjustment of pulmonary airflow and blood flow to each other.

15. Most CO_2 in the blood chemically reacts with water, especially in the RBCs, where there is an enzyme, ___, that catalyzes this reaction.

B. Matching

A. phrenic nerves	G. Charles' law	M. H_2CO_3	S. carboxyhemoglobin
B. larynx	H. nasal apertures	N. Boyle's law	T. inspiratory reserve volume
C. Bohr effect	I. hyperpnea	O. conchae	U. minute respiratory volume
D. trachea	J. atelectasis	P. pharynx	V. bronchoconstriction
E. cor pulmonale	K. hypercapnia	Q. pneumothorax	W. intercostal nerves
F. $H_2CO_3^- + H^+$	L. Dalton's law	R. vibrissae	X. carbaminohemoglobin

1. Right-heart failure due to pulmonary circulation obstruction
2. Explains why the heating of inhaled air aids inflation of the lungs
3. Contains the vocal cords
4. Tidal volume × respiratory rate
5. Stimulate the diaphragm to contract
6. Collapse of a lung
7. Additional air one can inhale beyond the tidal volume
8. Formed by the binding of CO_2 to hemoglobin
9. $CO_2 + H_2O \rightarrow$
10. Excessive P_{CO_2} in the blood

C. True or False

1. A hemoglobin molecule can transport O_2 and CO_2 simultaneously, but not O_2 and CO.

2. Atelectasis is always caused by pneumothorax.

3. When you hold your breath, the increasingly strong stimulus to breathe is the growing deficiency of oxygen in the blood.

4. When oxyhemoglobin passes through a systemic capillary, it normally unloads nearly 100% of its oxygen.

5. Hemoglobin unloads more oxygen at a low pH than it does at a higher one.

6. Hypoventilation can lead to hypercapnia and respiratory acidosis.

7. It requires more muscular effort to inhale than to exhale.

8. We cannot increase our blood P_{O_2} significantly by breathing more heavily.

9. Chronic bronchitis and emphysema are the major COPDs and are caused by allergens in the environment.

10. Restrictive disorders of respiration reduce pulmonary compliance and vital capacity.

D. Multiple Choice

1. The most abundant gas in the air you inhale is ___, and the most abundant gas in the air you exhale is ___. (a) oxygen, carbon dioxide; (b) oxygen, water vapor; (c) nitrogen, water vapor; (d) nitrogen, nitrogen; (e) nitrogen, carbon dioxide

2. The most powerful stimulus to breathe is: (a) the pH of the blood; (b) the P_{CO2} of the blood; (c) the P_{O2} of the blood; (d) the pH of the cerebrospinal fluid; (e) the P_{O_2} of the cerebrospinal fluid.

3. A change in lung volume relative to a specific pressure change: (a) results in a pneumothorax; (b) leads to chronic bronchitis; (c) is called pulmonary compliance; (d) can reverse pulmonary edema; (e) is an incurable condition.

4. ___ is rapid and deep in response to acidosis. (a) Hyperpnea; (b) Hyperventilation; (c) Tachypnea; (d) Dyspnea; (e) Kussmaul respiration

5. Most of the CO_2 transported in the blood is: (a) bound to the Fe^{2+} of hemoglobin; (b) dissolved gas bubbles; (c) in the form of carbonic acid and HCO_3^-; (d) bound to the globin moiety of hemoglobin; (e) bound to carbonic anhydrase.

6. According to Henry's law, the maximum amount of gas that will dissolve in a liquid depends on all of these factors *except:* (a) the temperature of the liquid; (b) the molecular weight of the gas; (c) the partial pressure of the gas; (d) the solubility of the gas; (e) all of these are important.

7. Which of these can occur when you exhale? (a) contraction of the scalenes; (b) doming up of the diaphragm; (c) relaxation of the rectus abdominis; (d) contraction of the external intercostals; (e) contraction of the erector spinae

8. A condition in which exhaling is more difficult than inhaling is: (a) emphysema; (b) tuberculosis; (c) pneumonia; (d) hypoxia; (e) pleurisy.

9. The nasal cavity opens anteriorly by the nostrils and posteriorly (into the pharynx) by the: (a) posterior nasal apertures; (b) vibrissae; (c) conchae; (d) fauces; (e) meatuses.

10. One of the potential effects of hypoxia is: (a) pneumothorax; (b) atelectasis; (c) low hematocrit; (d) pernicious anemia; (e) cyanosis.

11. According to ___, a tissue with a high metabolic rate will have a low pH, and this will induce hemoglobin to unload oxygen to it. (a) Poiseuille's law; (b) Dalton's law; (c) the Haldane effect; (d) the Bohr effect; (e) Henry's law

12. The amount of air you can inhale and exhale with the greatest possible effort, called the ___, is an important measure of your respiratory health. (a) tidal volume; (b) inspiratory reserve volume; (c) inspiratory capacity; (d) functional residual capacity; (e) vital capacity

13. The venous blood coming out of your quadriceps femoris right now probably has about ___ the blood going into it. (a) the same amount of CO_2 as; (b) 10–12% less CO_2 than; (c) 20–25% less O_2 than; (d) 65–72% less O_2 than; (e) 97% less O_2 than

14. In ___, the glottis is closed and the abdominal muscles are contracted to raise pressure in the abdominal cavity. (a) the Heimlich maneuver; (b) Valsalva's maneuver; (c) Henry's maneuver; (d) dyspnea; (e) eupnea

15. Emphysema reduces oxygen loading by the pulmonary blood because it: (a) reduces the surface area of the respiratory membrane; (b) reduces oxygen solubility; (c) reduces the concentration gradient of oxygen from air to blood; (d) thickens the respiratory membrane; (e) reduces pulmonary compliance.

16. All of these cause increased oxyhemoglobin dissociation *except:* (a) increased BPG;
    (b) decreased pH; (c) decreased temperature; (d) the Bohr effect; (e) thyroxine.

17. In a state of rest, the diaphragm moves about ___ during each inspiration. (a) 2–3 mm upward;
    (b) 15 mm downward; (c) 6–7 cm upward; (d) 10–15 cm downward; (e) 15–20 cm upward

18. Cessation of breathing for a few seconds during sleep is called: (a) apnea; (b) dyspnea;
    (c) pneumonia; (d) atelectasis; (e) eupnea.

19. Alveolar gas exchange depends on all *except:* (a) membrane thickness; (b) ventilation–perfusion
    coupling; (c) alveolar surface area; (d) concentration gradient of gases; (e) respiration rate.

20. Vital capacity is the sum of inspiratory reserve volume, expiratory reserve volume, and: (a) total
    lung capacity; (b) tidal volume; (c) residual volume; (d) functional residual capacity;
    (e) maximum voluntary expiration.

21. Which of these is/are true?
    1. An important function of the respiratory system is acid–base balance.
    2. There is a mucociliary escalator associated with the respiratory system that removes dust,
       pathogens and disease agents from the airways.
    3. Nasal conchae increase the surface area of the nasal cavity and warm, humidify, and clean the
       air we breathe.
    4. If one suffers from Ondine's curse, one cannot go to sleep.

    (a) 1 & 3; (b) 2 & 4; (c) 1, 2, & 3; (d) 4 only; (e) all the above

22. Which of these is/are true about blood gases?
    1. Decreased pH shifts the oxyhemoglobin dissociation curve to the right.
    2. In exercising skeletal muscle, HbO_2 dissociation is greater than it is in the same muscle
       at rest.
    3. The pressure gradient for O_2 between pulmonary blood and tissue fluid is steeper than that
       for CO_2.
    4. A low concentration of HbO_2 in the blood allows the blood to carry more CO_2.

    (a) 1 & 3; (b) 2 & 4; (c) 1, 2, & 3; (d) 4 only; (e) all the above

23. Which of these is/are true?
    1. The survival rate among lung cancer patients is good if the cancer is caught early.
    2. Receiving too much oxygen under pressure can damage the nervous system.
    3. Most lung tumors begin in the alveolar sacs.
    4. Acidosis causes an increased ventilation rate.

    (a) 1 & 3; (b) 2 & 4; (c) 1, 2, & 3; (d) 4 only; (e) all the above

24. Which of these is/are true about mechanisms of ventilation?
    1. When lung compliance decreases, lung inflation increases, according to Charles' law.
    2. The lungs, like the heart, have a pacemaker located in the brain stem.
    3. When intrapulmonary pressure rises, the chest wall moves up and out.
    4. When air is inhaled through the nose, its volume increases, and this helps inflate the lungs.
    (a) 1 & 3; (b) 2 & 4; (c) 1, 2, & 3; (d) 4 only; (e) all the above

25. Which of these is/are true?
    1. At high elevation, the P_{O2} is lower than at sea level. ✓
    2. Carbon monoxide competes with O_2-binding sites on hemoglobin and binds more tightly than does O_2.
    3. Both O_2 and CO_2 are carried dissolved in plasma and bound to hemoglobin.
    4. Because CO_2 is less soluble in water than O_2, there is an equal exchange across the alveolar surface of these two gases.

    (a) 1 & 3; (b) 2 & 4; (c) 1, 2, & 3; (d) 4 only; (e) all the above

E. Word Origins

1. In *nasal conchae, conch-* means "swirl, spiral."
2. In *cricoid, cric-* means "neck."
3. In *epiglottis, epi-* means "above, upon."
4. In *epiglottis, -glottis* means "throat."
5. In *corniculate, corni-* means "horn."
6. In *atelectasis, -ectasis* means "extension."
7. In *spirometer, spiro-* means "spiral, coil."
8. In *eupnea, eu-* means "normal, easy."
9. In *eupnea, -pnea* means "lung."
10. In *hypercapnia, capn-* means "smoke."
11. dys-
12. emphys-
13. -oma
14. aryten-
15. choana-

F. Which One Does Not Belong?

1. (a) respiratory bronchiole; (b) alveolar duct; (c) bronchiole; (d) alveolus

2. (a) pleurae; (b) trachea; (c) bronchioles; (d) larynx

3. (a) scalenes; (b) external intercostals; (c) rectus abdominis; (d) serratus anterior

4. (a) cricoid cartilage; (b) vestibular folds; (c) cuneiform cartilage; (d) alar cartilage

5. (a) inspiratory reserve volume; (b) tidal volume; (c) vital capacity; (d) expiratory reserve volume

G. Figure Exercise

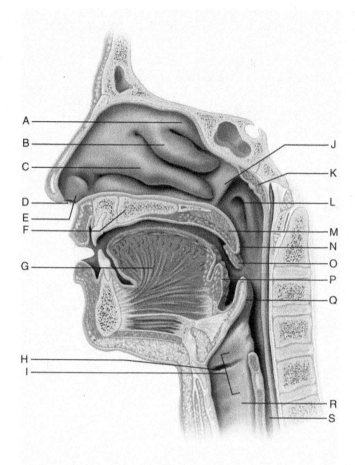

Match the statements with the structures in the diagram. Some answers may be used more than once; some questions can have more than one answer.

1. Structure(s) that assure(s) maximum contact with the air
2. Helps trap and destroy airborne pathogens larger than 10 μm
3. Most common site for epistaxis
4. Made of elastic cartilage; covers the laryngeal opening
5. Contains a venous plexus and erectile tissue that periodically restricts blood flow
6. Contains cricoid and thyroid cartilages
7. Contains olfactory cells
8. Contained in the oropharynx
9. Houses ciliated pseudostratified respiratory mucosa; ciliamobile
10. Contained in the nasopharynx

23   The Urinary System

A.   Short Answer

1. ___ is the concentration of solute particles in a solution, measured as moles of dissolved particles per liter of solution.

2. The major nitrogenous wastes in the blood are ___, ___, and creatinine.

3. All parts of a nephron, except for the nephron loops, are located in a layer of the kidney called the ___.

4. Arteries called the ___ ascend from the arcuate artery toward the kidney surface and give off one ___ to each nephron along the way.

5. Urine formation begins with the filtration of fluid from capillaries of the ___.

6. ___ stimulates the distal convoluted tubule and thick ascending limb of the nephron loop to increase calcium reabsorption.

7. Most tubular reabsorption occurs in the ___, the first duct distal to the glomerular capsule.

8. Cuboidal cells of the duct in question 7 have numerous long ___ on their luminal surfaces that allow extensive tubular reabsorption.

9. If a solute enters the duct in question 7 at a rate that exceeds the ___ for its reabsorption, the excess stays in the tubule and appears in the urine.

10. ___ can be an occupational hazard of truck drivers and motorcyclists; the kidneys slip below their normal position on the posterior abdominal wall.

11. The salinity gradient of the renal medulla is maintained by the ___, which is short or absent in aquatic animals that cannot conserve water.

12. The ___ apparatus monitors the salinity of the tubular fluid and adjusts the ___ rate to maintain homeostasis.

13. In response to low blood pressure, the kidneys secrete an enzyme called ___, which is the first step in the synthesis of a vasoconstrictor called ___.

14. The reabsorption of glucose by the proximal convoluted tubule depends on the simultaneous reabsorption of ___, since the SGLT protein binds and transports both of these at once.

15. The ___ sphincter is controlled by a parasympathetic reflex, whereas the ___ sphincter is under voluntary control.

B. Matching

A. vasa recta	G. urea	M. creatinine	S. polyuria
B. alcohol	H. specific gravity	N. nephron loop	T. furosemide
C. osmolality	I. detrusor	O. macula densa	U. external urinary sphincter
D. bulbospongiosus	J. caffeine	P. osmolarity	V. collecting duct
E. hematuria	K. azotemia	Q. renal clearance	W. cortical radiate artery
F. uremia	L. arcuate artery	R. carbon dioxide	X. glomerular filtration rate

1. Metabolic waste that is not a nitrogenous waste
2. Consequence of azotemia
3. Detects flow rate and fluid composition of the distal convoluted tubule
4. Consequence of many forms of diabetes
5. The countercurrent multiplier
6. Primary site of ADH action
7. Muscle of the bladder wall
8. Diuretic that impairs countercurrent multiplier function
9. Net filtration pressure × filtration coefficient
10. Accumulation of nitrogenous wastes in the blood   *azotemis*

C. True or False

1. Osmotic and electrical gradients created by sodium reabsorption in the proximal convoluted tubule facilitate reabsorption of other solutes.

2. Aldosterone is secreted by the granular cells of the juxtaglomerular apparatus.

3. All urea is cleared from the bloodstream as blood passes through the kidney, so blood leaving the renal vein is urea-free.

4. Each collecting duct of the kidney drains several nephrons.

5. GFR rises if the efferent arteriole constricts and other factors remain the same.

6. No matter how dehydrated one becomes, it is impossible to excrete less than 400 mL of urine per day.

7. Aldosterone has no effect on how much sodium the proximal convoluted tubule reabsorbs.

8. If T_{max} for glucose is exceeded at the filtration membrane, glucose acts as an osmotic diuretic in the nephron.

9. Two percent NaCl and 2% sucrose have the same osmolarity.

10. Except in infants and incontinent adults, micturition does not occur until one voluntarily contracts the detrusor muscle.

E gav

D. Multiple Choice

1. Podocytes are located in the: (a) cuboidal cells of the proximal convoluted tubule; (b) visceral layer of the glomerular capsule; (c) thin limb of the nephron loop; (d) juxtaglomerular apparatus; (e) renal pyramids.

2. A kidney has more ___ than it has of any of these other structures. (a) afferent arterioles; (b) collecting ducts; (c) arcuate arteries; (d) cortical radiate arteries; (e) renal pyramids

3. The ___ supply blood to the renal medulla, but they are arranged in such a way that they do not remove salt from the tissue. (a) vasa recta; (b) vasa efferentia; (c) peritubular capillaries; (d) arcuate arteries; (e) cortical radiate arteries

4. Parathyroid hormone: (a) acts on the PCT and stimulates calcium reabsorption; (b) stimulates the PCT to increase phosphate excretion; (c) is an effective osmoregulator; (d) causes calcium excretion in the CD; (e) stimulates calcitonin release.

5. As a molecule of urea passes from the bloodstream to the tubular fluid, the first thing it passes through is: (a) a macula densa; (b) a filtration slit; (c) the parietal wall of the glomerular capsule; (d) the basement membrane of the capillary; (e) a fenestration in a glomerular capillary.

6. When glucose is found in the urine of a patient, you know that: (a) nitrogenous wastes are not being secreted; (b) the filtration barrier is normal; (c) the transport maximum for glucose has been exceeded; (d) the renin–angiotensin–aldosterone system has failed; (e) the patient has diabetes insipidus.

7. Blood pressure is unusually high in the glomerulus because the ___ is smaller than the ___. (a) efferent arteriole, peritubular capillaries; (b) descending limb, ascending limb; (c) efferent arteriole, afferent arteriole; (d) proximal convoluted tubule, distal convoluted tubule; (e) cortical radiate artery, cortical radiate vein

8. Some animals cannot produce hypertonic urine because they do not have: (a) glomeruli; (b) glomerular capsules; (c) aldosterone; (d) nephron loops; (e) proximal convoluted tubules.

9. About 65% of the fluid filtered by the glomerulus is reabsorbed by the: (a) proximal convoluted tubule; (b) nephron loop; (c) distal convoluted tubule; (d) collecting duct; (e) calices.

10. The daily output of urine is about ___ of the fluid that the kidneys initially filter from the bloodstream. (a) 1–2%; (b) 5–12%; (c) 25%; (d) 50%; (e) 90–95%

11. Renin functions by: (a) stimulating the adrenal glands to release aldosterone; (b) targeting the glomerular capsule; (c) converting angiotensinogen to angiotensin I; (d) stimulating the macula densa to produce EPO; (e) causing the mesangial cells to deliver more NaCl to the collecting duct.

12. All of these are functions of the kidneys *except:* (a) electrolyte balance; (b) urea production; (c) gluconeogenesis; (d) blood pressure regulation; (e) clearance of certain drugs.

13. In a juxtamedullary nephron, the thick ascending segment of the nephron loop is permeable to ___ but not ___; therefore, the tubular fluid becomes more and more dilute as it flows along this segment. (a) water, NaCl; (b) urea, NaCl; (c) Na^+, water; (d) Na^+, K^+; (e) NaCl, water

14. The water permeability of the collecting duct is influenced by: (a) the GFR; (b) atrial natriuretic peptide; (c) angiotensin II; (d) aldosterone; (e) antidiuretic hormone.

15. Which of the following would reduce the GFR? (a) constriction of the afferent arteriole; (b) constriction of the efferent arteriole; (c) dilation of the afferent arteriole; (d) a drop in oncotic pressure; (e) aldosterone hypersecretion

16. The most selective part of the filtration membrane in the glomerulus is the: (a) glomerular capillary membrane; (b) basement membrane; (c) pedicles; (d) podocytes; (e) filtration slits.

17. The renal pelvis receives urine from: (a) the glomerulus; (b) collecting ducts; (c) the ureters; (d) the minor calices; (e) the major calices.

18. Which of these is not a step in urine formation? (a) tubular secretion; (b) glomerular filtration; (c) tubular excretion; (d) water conservation; (e) tubular reabsorption

19. The function of the juxtaglomerular apparatus is to: (a) concentrate urine; (b) regulate the GFR; (c) regulate urine pH; (d) reabsorb glucose; (e) maintain the salinity of the medulla.

20. ACE finishes a job begun by: (a) renin; (b) angiotensin II; (c) aldosterone; (d) atrial natriuretic peptide; (e) erythropoietin.

21. Glomerular filtration rate is controlled by:
    1. autoregulation mechanisms, including juxtaglomerular feedback.
    2. parasympathetic stimulation.
    3. activation of the renin–angiotensin–aldosterone mechanism.
    4. increasing reabsorption rate.
    (a) 1 & 3; (b) 2 & 4; (c) 1, 2, & 3; (d) 4 only; (e) all the above

22. Which of these is/are true?
    1. Nephrosclerosis is one result of hypertension.
    2. Creatinine, once filtered, is excreted in urine.
    3. Intercalated cells of the nephron function in acid–base regulation.
    4. Beavers and other aquatic mammals, such as kangaroo rats, have no juxtamedullary nephrons and, therefore, cannot conserve water.

    (a) 1 & 3; (b) 2 & 4; (c) 1, 2, & 3; (d) 4 only; (e) all the above

23. Which of these is/are true?
    1. BUN is a measure of GFR.
    2. Both the PCT and DCT have cuboidal cells with extensive brush borders.
    3. Blood proteins normally pass through the filtration membrane into the glomerular filtrate.
    4. Urea is a waste product of protein metabolism.

    (a) 1 & 3; (b) 2 & 4; (c) 1, 2, & 3; (d) 4 only; (e) all the above

24. Which is/are true about the function of parts of the nephron?
    1. Tubular secretion occurs in the PCT, DCT, and nephron loop.
    2. In the PCT, Na^+ and some nutrients are reabsorbed by the transcellular route.
    3. EPI and NE target the JGA and afferent arteriole.
    4. Obligatory water reabsorption takes place in the PCT.

    (a) 1 & 3; (b) 2 & 4; (c) 1, 2, & 3; (d) 4 only; (e) all the above

25. Which of these is/are true?
    1. Both NaCl and urea add to the solute gradient in the inner medullary region.
    2. The proper function of the countercurrent multiplier depends on fluid flowing in opposite directions in the two limbs of the loop.
    3. The countercurrent exchanger, the vasa recta, removes water from the medulla that the nephron reclaims from the tubular fluid.
    4. Under the influence of ADH, the collecting duct reabsorbs more water, making the urine less concentrated.

    (a) 1 & 3; (b) 2 & 4; (c) 1, 2, & 3; (d) 4 only; (e) all the above

E. Word Origins

1. In *renal, ren-* means "urinary."
2. In *renal calyx, calyx* means "outer layer."
3. In *hydronephrosis, nephr-* means "kidney."
4. In *glomerulus, glomer-* means "ball."
5. In *podocyte, podo-* means "porous."
6. In *pedicel, pedi-* means "foot."
7. In *macula densa, macula* means "patch, spot."
8. In *oliguria, olig-* means "too much."
9. In *cystitis, cyst-* means "bladder."
10. In *insipidus, -insipid* means "infection."
11. -tripsy
12. angi-
13. juxta-
14. melli-
15. mictur-

F. Which One Does Not Belong?

1. (a) aldosterone; (b) ADH; (c) ANP; (d) angiotensin II

2. (a) basement membrane; (b) filtration slits; (c) glomerular capillary membrane; (d) mesangial cells

3. (a) renin; (b) erythropoietin; (c) calcitriol; (d) urea

4. (a) collecting duct; (b) DCT; (c) PCT; (d) nephron loop

5. (a) endocrine system; (b) respiratory system; (c) integumentary system; (d) urinary system

G. Figure Exercise

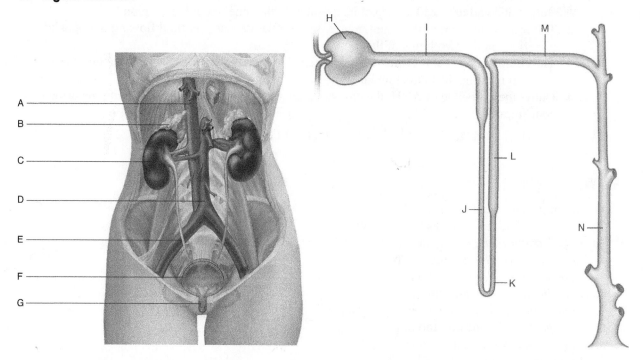

Match the statements with the structures in the diagram. Some answers may be used more than once; some questions can have more than one answer.

1. Takes urine from renal pelvis to bladder
2. Produce(s) hormones
3. Site where PTH stimulates phosphate excretion and calcitriol synthesis
4. Generate(s) salinity gradient for urine concentration
5. Component(s) of countercurrent multiplier
6. Target for a hormone produced in the heart
7. Its cells have nuclear receptors for end product of renin–angiotensin mechanism
8. Target for protein hormone that conserves water
9. Area(s) most permeable to urea
10. Houses podocytes
11. Contains brush border cells
12. Has principal cells with hormone receptors and intercalated cells for acid–base balance
13. Reabsorption site for most nutrients filtered
14. Contains macula densa
15. Impermeable to water

24 Water, Electrolyte, and Acid–Base Balance

A. Short Answer

1. ___ water loss is a loss of which we are usually unaware, such as water lost in the expired breath.

2. A deficiency of sodium in the blood is called ___.

3. A(n) ___ imbalance is less critical than an acid–base imbalance.

4. A blood pH above 7.45 is called ___.

5. A pH imbalance is said to be ___ if the body's buffer systems cannot restore acid–base homeostasis.

6. Neurons of the hypothalamus that monitor the osmolarity of the blood are called ___.

7. A(n) ___ is an acid that ionizes to only a slight extent in solution, so it does not greatly alter the pH.

8. The ICF and ECF are the body's two major ___.

9. Water is lost through the skin not only by the secretion of sweat but also by a diffusion process called ___.

10. Hyperkalemia can cause acidosis because K^+ enters cells and drives ___ from the ICF into the ECF.

11. A fluid imbalance called ___ can result when the body loses more Na^+ than water—for example, when you sweat heavily and replace this fluid by drinking plain water.

12. ___ stimulates cells of the distal convoluted tubule to reabsorb Na^+.

13. A chemical mixture that converts a strong acid to a weak acid and, thus, minimizes the effects of the strong acid on the pH of the solution is called a(n) ___.

14. In acidosis, cells of the renal tubules catabolize amino acids and secrete ___ into the tubular fluid to buffer the excess acid.

15. Cells of the renal collecting ducts install proteins called aquaporins in their plasma membranes in response to the hormone ___.

B. Matching

A. edema	G. 7.35	M. hypocalcemia	S. hypotonic hydration
B. hypokalemia	H. 6.8	N. volume excess	T. juxtaglomerular cells
C. dehydration	I. 4.5	O. volume depletion	U. sequestration
D. adrenal cortex	J. Na^+	P. tissue fluid	V. posterior pituitary
E. hypercalcemia	K. K^+	Q. hypernatremia	W. respiratory acidosis
F. Cl^-	L. Ca^{+2}	R. hypochloremia	X. respiratory alkalosis

  1. Excess fluid collects in specific locations
  2. Deficiency of total body water, but with normal osmolarity
  3. Storage and release site for ADH
  4. Primary ICF ion
  5. Accumulation of excess fluid in interstitial spaces
  6. Most abundant cation of the ECF
  7. Source of aldosterone
  8. Effect of hypoparathyroidism
  9. Limiting pH of the renal tubular fluid
10. pH imbalance that may result from a degenerative lung disease

C. True or False

1. A person with diabetic acidosis is likely to have more ammonium chloride in the urine than a healthy person.

2. Hyperkalemia can cause acidosis, and acidosis can cause hyperkalemia.

3. Aldosterone substantially increases the sodium concentration of the ECF.

4. Excessive vomiting for several days can cause metabolic acidosis.

5. pH values up to 7.00 are called acidic, and those above 7.00 are alkaline. A blood pH of 7.2 is, therefore, considered to be alkalosis.

6. The sense of thirst is briefly satisfied just by moistening and cooling the mouth and filling the stomach, but long-term satiation depends on hydration of the blood.

7. The body's only sources of water are food and drink.

8. The kidneys can compensate for fluid excess much better than they can compensate for fluid deficiency.

9. Hyperkalemia is more life-threatening than hypokalemia.

10. When the tissue fluids are hypertonic, the kidney tubules reabsorb water by active transport.

D. Multiple Choice

1. If the body contains too much fluid but the fluid is isotonic, a person has: (a) hypovolemia; (b) fluid sequestration; (c) negative water balance; (d) positive water balance; (e) volume excess.

2. Which of the following is the *least* likely consequence of edema? (a) tissue necrosis; (b) elevated blood pressure; (c) suffocation; (d) nausea; (e) circulatory shock

3. Over the course of an average day, the body loses the least amount of water by way of: (a) urine; (b) feces; (c) expired breath; (d) sweat; (e) cutaneous transpiration.

4. All of the following can cause dehydration *except:* (a) overuse of diuretics; (b) aldosterone hypersecretion; (c) profuse sweating; (d) diabetes insipidus; (e) exposure to cold temperatures.

5. When hyperkalemia occurs quickly, such as in a crush injury: (a) cells become hyperpolarized; (b) vomiting ensues; (c) aldosterone secretion ceases; (d) nerve and muscle cells become more excitable; (e) hyperparathyroidism follows.

6. All of these stimulate aldosterone secretion *except:* (a) angiotensin II; (b) hyperkalemia; (c) hypercalcemia; (d) hypotension; (e) hyponatremia.

7. Volume excess (hypervolemia) can lead to acidosis because it: (a) depresses respiration; (b) causes oliguria; (c) promotes renal H^+ secretion; (d) reduces renal H^+ secretion; (e) promotes renal bicarbonate reabsorption.

8. Aldosterone acts as the "salt-retaining hormone" by: (a) stimulating water retention by the kidney; (b) increasing the number of Na^+–K^+ pumps in some regions of the nephron; (c) causing renin release from the JG apparatus; (d) inhibiting ANP; (e) stimulating osmoreceptors in the hypothalamus.

9. Which of the following has/have the greatest effect in buffering the pH of the blood plasma? (a) proteins; (b) bicarbonate; (c) chloride; (d) ammonia; (e) phosphates

10. The normal pH of the arterial blood is ___ ± 0.05. (a) 7.3; (b) 6.4; (c) 7.0; (d) 7.4; (e) 8.2

11. Most H^+ secreted into the tubular fluid eventually passes into the urine in the form of: (a) HCO_3^-; (b) H_2O; (c) NH_3; (d) carbonic anhydrase; (e) urea.

12. Ketones in the blood may cause: (a) diabetes mellitus; (b) metabolic alkalosis; (c) metabolic acidosis; (d) respiratory alkalosis; (e) respiratory acidosis.

13. After several days of severe vomiting, a pregnant woman suffers from alkalosis. We can predict that her urine will have an elevated concentration of: (a) ammonia; (b) urea; (c) bicarbonate; (d) hydrogen ions; (e) sodium dihydrogen phosphate.

14. To secrete a hydrogen ion into the tubular fluid, a kidney tubule must reabsorb a(n) ___ ion from it. (a) phosphate; (b) sodium; (c) potassium; (d) ammonium; (e) hydroxyl

15. Hypoventilation will produce acidosis because: (a) CO_2 accumulates and reacts with water to form carbonic acid which ionizes, releasing H^+; (b) H^+ is not expelled by the lungs as fast as it is produced; (c) excess OH^- is lost in the urine; (d) HCO_3^- is lost in the urine; (e) the phosphate buffer system cannot work effectively under hypoxic conditions.

16. Which of these is not regarded as a transcellular fluid? (a) lymph; (b) synovial fluid; (c) pleural fluid; (d) aqueous humor; (e) cerebrospinal fluid

17. Aquaporins are found in the: (a) glomerular capillaries; (b) proximal convoluted tubule; (c) parietal wall of the glomerular capsule; (d) collecting duct; (e) nephron loop.

18. True dehydration exists when: (a) daily water output is less then 400 mL; (b) total body water is low and the fluid is hypertonic; (c) proportionate amounts of water and salt are lost from the body; (d) total body water is low and the fluid is hypotonic; (e) total body water is normal but its osmolarity is abnormally high.

19. Hypovolemia promotes alkalosis because it: (a) promotes the reabsorption of H^+; (b) promotes the secretion of H^+; (c) dilutes the blood bicarbonate; (d) drives H^+ out of cells; (e) drives H^+ into cells.

20. The kidneys cannot secrete any more H^+ if: (a) there is no more bicarbonate in the blood; (b) there is no more bicarbonate in the tubular fluid; (c) the pH of the tubular fluid is below 4.5; (d) the pH of the tubular fluid is above 8.2; (e) there is too much ammonia in the tubular fluid.

21. Which of these is/are true?
    1. Prolonged exposure to cold can inhibit ADH secretion.
    2. Hypovolemia can be brought on by diabetes insipidus.
    3. Dehydration can lead to circulatory shock.
    4. Adults tend to be more prone to dehydration than children because of their larger size.

    (a) 1 & 3; (b) 2 & 4; (c) 1, 2, & 3; (d) 4 only; (e) all the above

22. Which of these is/are true about buffer systems?
    1. When kidneys excrete bicarbonate ions, the plasma pH decreases.
    2. The kidneys excrete hydrogen ions bound to bicarbonate, phosphate, and ammonia buffers.
    3. The respiratory system is more efficient in regulating pH than is the chemical buffer system.
    4. Proteins are the best buffer system in the ICF.

    (a) 1 & 3; (b) 2 & 4; (c) 1, 2, & 3; (d) 4 only; (e) all the above

23. Which of these is/are true?
    1. ADH can change plasma sodium concentration, but aldosterone does not.
    2. Estrogen and progesterone stimulate natriuresis.
    3. A slow increase in ECF potassium makes muscle and nerve cells less likely to depolarize.
    4. Hypernatremia is a result of drinking too much water.

    (a) 1 & 3; (b) 2 & 4; (c) 1, 2, & 3; (d) 4 only; (e) all the above

24. Which of these is/are true about fluid compartments?
   1. About 25% of total body water is found in the ECF.
   2. The ICF contains the largest amount of body water.
   3. Water distribution is independent of age and gender.
   4. Water must pass through membranes to go from one compartment to another, and its movement is determined by the solute concentration of adjacent compartments.

   (a) 1 & 3; (b) 2 & 4; (c) 1, 2, & 3; (d) 4 only; (e) all the above

25. Which of these is/are true?
   1. Sodium reabsorption from the nephron alters plasma osmolarity even when ADH is present.
   2. Urine concentration of 90 mOsm means that ADH secretion has increased.
   3. Aquaporins increase in the collecting duct under the influence of angiotensin II.
   4. The osmoreceptors in the hypothalamus are sensitive to ADH, angiotensin II, and the osmolarity of the ECF.

   (a) 1 & 3; (b) 2 & 4; (c) 1, 2, & 3; (d) 4 only; (e) all the above

E. Word Origins

1. In *transpiration, trans-* means "temporary."
2. In *hypovolemia, hypo-* means "below normal."
3. In *hypovolemia, -emia* means "blood condition."
4. In *sequestration, sequestr-* means "to hide."
5. In *hypernatremia, natr-* means "sodium."
6. In *hyperkalemia, kal-* means "calcium."
7. In *intracellular, intra-* means "within, inside."
8. In *extracellular, extra-* means "more than usual."
9. In *acidosis, -osis* means "normal."
10. In *parenteric, par-* means "beside."
11. vol-
12. spir-
13. calc-
14. enter-
15. aliment-

F. Which One Does Not Belong?

1. (a) interstitial fluid; (b) intracellular fluid; (c) plasma; (d) transcellular fluid

2. (a) hyperkalemia; (b) hypovolemia; (c) hypervolemia; (d) hyperchloremia

3. (a) Ca^{+2}; (b) Cl^-; (c) Na^+; (d) K^+

4. (a) hyponatremia; (b) hypotension; (c) hyperkalemia; (d) hyperphosphatemia

5. (a) vitamin D deficiency; (b) pregnancy; (c) hyperparathyroidism; (d) lactation

25 The Digestive System

A. Short Answer

1. Food is ground between the ___ surfaces of the premolars and molars.

2. Saliva contains salivary *Amylase*, which digests starch in the mouth, and lingual *Lipase*, which digests fat in the stomach.

3. Glands in the *Cardiac* ___ and *pyloric* ___ regions of the stomach secrete mainly mucus.

4. Bile acids are reused during digestion by means of secretion, reabsorption, and resecretion via a route called ___. (two words)

5. In the secretion of hydrochloric acid into the stomach, H^+ is exchanged for ___ from the lumen of gastric glands and Cl^- is exchanged for ___ from the blood.

6. In the ___ phase of gastric control, even the taste and sight of food can stimulate gastric secretion and contraction.

7. Peptic ulcers are caused by ___.

8. Most liver functions such as nutrient absorption, glycogen storage, and protein secretion are carried out by its cuboidal epithelial cells, the ___.

9. Enterokinase of the intestinal mucosa activates the enzyme ___, which converts other pancreatic zymogens into active enzymes.

10. ___ stimulates the pancreas to secrete enzymes, initiates gallbladder contraction, and causes the hepatopancreatic sphincter to relax.

11. The ___ reflex is a mechanism that promotes emptying of the large intestine (defecation) when chyme enters the duodenum.

12. The first enzyme in the GI tract that begins to digest dietary protein is ___.

13. A pancreatic enzyme that removes amino acids one at a time from the –COOH end of a peptide is ___.

14. Absorptive cells of the small intestine package digested lipids with proteins in droplets called ___ that can be transported in the lymph and blood.

15. Overlapping waves of peristalsis in the small intestine are called the ___.

B. Matching

A. canines	G. fauces	M. monosaccharides	S. myenteric plexus
B. amylase	H. lacteals ~	N. premolars	T. taeniae coli
C. chymotrypsin	I. pepsin	O. secretin	U. intestinal crypts
D. amino acids	J. micelles	P. gastrin	V. vomiting
E. swallowing	K. submucosal plexus	Q. submucosa	W. haustral contractions
F. frenulum	L. intrinsic factor	R. villi	X. molars /

1. Adults have only four *Canines*
2. Function of the emetic center of the medulla oblongata *Swallowing*
3. The only indispensable secretion of the stomach *Intrinsic factor*
4. Muscularis externa of the large intestine *Taeniae Coli*
5. Anchors each lip to the gum as well as the tongue to the floor of the mouth *Frenulum*
6. Lymph capillaries in intestinal villi *lacteals*
7. Stimulates the pancreas and liver to secrete sodium bicarbonate *Secretin*
8. Digests protein in the small intestine *Chymotrypsin*
9. Stimulates secretion of HCl from the stomach *Gastrin*
10. Structure present in the small intestine but not in the large intestine *Villi*

C. True or False

1. All dietary lipids circulate to the liver before going to any other organ. *F*

2. Only a small portion of calcium in green leafy vegetables is absorbed because it is bound to oxalate, preventing its absorption.

3. Bile salts enable pancreatic lipase to work more efficiently.

4. Vitamins and minerals are not digested.

5. Intrinsic factor is necessary for the absorption of calcium by the small intestine. *B¹²*

6. Dietary proteins cannot be absorbed into the bloodstream without first being hydrolyzed completely to amino acids. *Some proteins are absorbed by pinocytosis.*

7. Most cells of the intestinal crypts are the type called chief cells. *stomach*

8. Secretin reacts with HCl and neutralizes it in the small intestine. *Sodium bicarbonate neutralizes HCL*

9. Pepsin produces more pepsin.

10. The liver does not secrete any digestive enzymes. *T*

D. Multiple Choice

1. Which of the following terms is most nearly equivalent to digestion? (a) anabolism; (b) polymerization; (c) condensation; (d) hydrolysis; (e) assimilation

2. The myenteric nerve plexus is located between the layers of the: (a) greater and lesser omenta; (b) lamina propria and submucosa; (c) muscularis externa; (d) muscularis mucosae and muscularis externa; (e) muscularis mucosae and submucosa.

3. Blood vessels and bile passages enter and leave the liver at a point on its inferior surface called the: (a) porta hepatis; (b) antrum; (c) pylorus; (d) cecum; (e) cervix.

4. The ___ regulates the passage of chyme from the stomach to the duodenum.
    (a) lower esophageal sphincter; (b) antrum; (c) ileocecal valve; (d) greater duodenal papilla; (e) pyloric sphincter

5. Hydrochloric acid is secreted by: (a) goblet cells; (b) parietal cells; (c) alpha cells; (d) chief cells; (e) G cells.

6. The intestinal crypts are most like the ___ in function. (a) villi; (b) lacteals; (c) teniae coli; (d) pyloric glands; (e) haustra

7. The function of lacteals is to: (a) absorb lipid-soluble products of digestion; (b) secrete mucus; (c) secrete digestive enzymes; (d) produce enteric hormones; (e) absorb chylomicrons.

8. The effect of bile salts on dietary fats is best described as: (a) lipolysis; (b) lipogenesis; (c) hydrolysis; (d) emulsification; (e) absorption.

9. The pancreas produces all of the following *except:* (a) pepsin; (b) trypsinogen; (c) chymotrypsinogen; (d) sodium bicarbonate; (e) ribonuclease.

10. The brown color of feces is produced by: (a) bilirubin; (b) bacteria; (c) urobilinogen; (d) meat in the diet; (e) indigestible material in food.

11. ___ are produced by the complete digestion of starch and protein. (a) Peptides; (b) Zymogens; (c) Chylomicrons; (d) Micelles; (e) Monomers

12. Cholecystokinin: (a) stimulates the appetite; (b) stimulates HCl secretion; (c) inhibits gastric secretion and motility; (d) relaxes the ileocecal valve; (e) stimulates mass movements of the colon.

13. The G cells of the stomach secrete: (a) GIP; (b) glucagon; (c) HCl; (d) gastrin; (e) pepsin.

14. Amylase digests starch to: (a) maltose and oligosacchrides; (b) polysaccharides; (c) oligopeptides; (d) maltase; (e) glucose.

15. ___ are droplets of dietary fat coated with bile salts and lecithin. (a) Micelles; (b) Chylomicrons; (c) Globules; (d) Emulsification droplets; (e) Adipocytes

16. The decidual dentition lacks: (a) enamel; (b) dentin; (c) premolars; (d) molars; (e) canines.
    ✓ milk teeth

17. The place occupied by cementum in the root of a tooth is occupied by ___ in the crown. (a) pulp; (b) enamel; (c) gingiva; (d) periodontal membrane; (e) dentin

18. Saliva does *not* contain: (a) electrolytes; (b) antibodies; (c) amylase; (d) mucus; (e) pepsin.

19. In the wall of the digestive tract, the ___ is immediately deep to the mucosal epithelium.
    (a) lamina propria; (b) muscularis externa; (c) serosa; (d) adventitia; (e) submucosa

20. Contact digestion is promoted especially by which type of gastrointestinal motility?
    (a) swallowing; (b) micturition; (c) haustral contractions; (d) segmentation; (e) peristalsis

21. Which of these is/are true about the stomach?
 1. Blood leaving it has a low pH because of the large amounts of HCl produced there.
 2. Chief cells produce the zymogen pepsinogen.
 3. Hydrochloric acid is used for protein digestion in the stomach.
 4. When food enters the stomach, its muscular wall relaxes reflexively.

   (a) 1 & 3; (b) 2 & 4; (c) 1, 2, & 3; (d) 4 only; (e) all the above

22. Which of these is/are true about the duodenum?
 1. It receives products made in the liver and pancreas via the hepatopancreatic sphincter.
 2. It secretes CCK, GIP, and secretin.
 3. Pancreatic zymogens are converted into active enzymes here.
 4. Hydrochloric acid from the stomach is neutralized here by bicarbonate from the pancreas.

   (a) 1 & 3; (b) 2 & 4; (c) 1, 2, & 3; (d) 4 only; (e) all the above

23. Which of these is/are true about the digestive system?
 1. As a result of the enterogastric reflex, the stomach is inhibited.
 2. Biliary calculi are most often caused by a high-cholesterol diet.
 3. No adult mammals drink milk, except some *Homo sapiens*.
 4. The stomach contains pacemaker cells that regulate its contractions.

   (a) 1 & 3; (b) 2 & 4; (c) 1, 2, & 3; (d) 4 only; (e) all the above

24. True statements about the liver include:
 1. hepatic macrophages (Kupffer cells) secrete many liver proteins.
 2. blood in liver sinusoids contains products of protein and carbohydrate digestion.
 3. bile is produced in the bile duct but then travels to the gallbladder.
 4. it degrades hormones, detoxifies the blood, and produces many blood proteins.

   (a) 1 & 3; (b) 2 & 4; (c) 1, 2, & 3; (d) 4 only; (e) all the above

25. Features of the small intestine include:
 1. villi, which house microvilli that increase the surface area of the mucosa.
 2. pacemaker cells, which regulate the rhythm of segmentation.
 3. intestinal crypts (of Lieberkühn) between villi.
 4. lymphatic nodules called Peyer's patches.

   (a) 1 & 3; (b) 2 & 4; (c) 1, 2, & 3; (d) 4 only; (e) all the above

Cheif Cells or Zymogenic cells
releases pepsinogen
& chymosin

E. Word Origins

1. In *alimentary, aliment-* means "digestion."
2. In *labial, labi-* means "lip."
3. In *parotid, par-* means "salivary."
4. In *mesentery, enter-* means "inside."
5. In *gastrointestinal, gastr-* means "stomach."
6. In *falciform, falci-* means "false."
7. In *sigmoid, sigm-* means "S-shaped."
8. In *rectum, rect-* means "straight."
9. In *micelle, -elle* means "little."
10. In *chylomicron, chylo-* means "paste."
11. haustr-
12. jejun-
13. emet-
14. pylorus
15. caud-

F. Which One Does Not Belong?

1. (a) stomach; (b) pancreas; (c) rectum; (d) duodenum

2. (a) alkaline tide; (b) chloride shift; (c) H^+–K^+ ATPase; (d) secretin

3. (a) sympathetic division; (b) myenteric plexus; (c) enteric nervous system; (d) mucosal plexus

4. (a) chylomicrons; (b) haustra; (c) micelles; (d) emulsification droplets

5. (a) submandibular gland; (b) buccal gland; (c) sublingual gland; (d) parotid gland

G. Figure Exercise

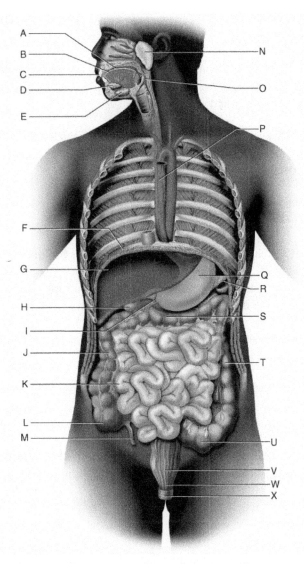

Match the statements with the structures in the diagram. Some answers may be used more than once; some questions can have more than one answer.

1. Produce(s) bile in canaliculi
2. Secretes hormone that stimulates HCl production
3. Used in mechanical digestion/chewing
4. Ends in cardiac orifice
5. Produce(s) mucus, lipase, and amylase
6. Most common movement is segmentation
7. Has rugae and three layers of smooth muscle
8. Area(s) required for swallowing
9. Site(s) of vitamin K synthesis
10. Contain(s) haustra
11. Produce(s) zymogens and bicarbonate ions
12. Houses hemorrhoidal veins
13. Site of most digestion and absorption of nutrients
14. Source of immune cells; populated with numerous lymphocytes
15. Concentrates bile

26   Nutrition and Metabolism

A. Short Answer

1. A loss of appetite, or ___, can result from destruction of the ___ of the hypothalamus.

2. A deficiency of blood glucose is called ___.

3. Glycogen synthesis is stimulated by ___, while glycogenolysis is triggered by ____ and ____.

4. The ___ in the brain is a center for appetite regulation.

5. ___ are bloodborne protein–lipid complexes that transport excess cholesterol and phospholipids to the liver for disposal.

6. A ___ protein is one that provides all eight ___ amino acids for building human proteins.

7. When muscle is broken down, as in starvation or muscular dystrophy, a person may have more nitrogen in the urine than in the diet and is said to be in a state of ___.

8. For the first few hours after a meal, a person is in a(n) ___ state, in which digested nutrients are being taken up by the blood.

9. The body reduces its ____ in order to conserve body mass, making losing weight more difficult.

10. Glycolysis decomposes glucose into two molecules of ___ as its end products.

11. When the diet is too low in carbohydrates, fats cannot be completely oxidized because there is not enough ___ to accept their C_2 breakdown products and get them into the citric acid cycle.

12. The postabsorptive state is primarily controlled by the ___ nervous system and by the hormone ____.

13. Using the ___ mechanism, mitochondria synthesize ATP by tapping energy from the H^+ gradient between the matrix and intermembrane space.

14. White foods have a high ___; they raise insulin demand and therefore increase the risk of type 2 diabetes mellitus.

15. ___ is a syndrome of cramps, vomiting, and hypotension that can result from extreme electrolyte loss in hot weather.

B. Matching

A. cellulose
B. NADH + FADH$_2$
C. ATP
D. hyperphagia
E. pectin
F. CO$_2$

G. ATP synthase
H. micronutrients
I. coenzyme A
J. cytochromes
K. D and K
L. B$_{12}$ and C

M. anorexia
N. gastric mucosa
O. proton pumps
P. coenzyme Q
Q. phosphorylation
R. Peptide YY

S. leptin
T. hemicellulose
U. pyruvic acid
V. lipoprotein lipase
W. macronutrients
X. HDL

1. Satiety signal from enteroendocrine cells of the ileum
2. Effect of damage to the satiety center
3. Vitamins and minerals
4. Cholesterol-lowering soluble fiber
5. Removes the fats from LDLs
6. Fat-soluble vitamins
7. First step of glycolysis
8. Most important product(s) of the citric acid cycle
9. Shuttles electrons from enzyme complex 1 to complex 2 in the mitochondria
10. Channel for H$^+$ diffusing back into the mitochondrial matrix

C. True or False

1. Water is not considered to be a nutrient.

2. Cellulose is an important nutrient, even though we cannot digest it.

3. Even in anaerobic fermentation, ATP synthesis occurs in the mitochondria.

4. In the equation for aerobic respiration, $C_6H_{12}O_6 + 6\ O_2 \rightarrow 6\ CO_2 + 6\ H_2O$, the water is produced by the mitochondrial electron transport chain, and the carbon dioxide is produced by the matrix reactions.

5. In spite of media reports of "good and bad cholesterol," cholesterol exists in only one form.

6. Glycogenesis breaks down stored glycogen and releases glucose into circulation.

7. High-density lipoproteins do not contribute to atherosclerosis.

8. More ATP is produced by the citric acid cycle than by any other stage of glucose catabolism.

9. The liver is responsible for most of the body's beta-oxidation of fatty acids.

10. When glucose is completely oxidized to CO_2 and H_2O, less than half of its energy is transferred to ATP.

D. Multiple Choice

1. Low-density lipoproteins contain the most ___ in comparison with other lipoproteins.
   (a) triglycerides; (b) cholesterol; (c) protein; (d) phospholipid; (e) glycogen

2. ___ are inorganic elements extracted by plants from soil and water. (a) Vitamins;
   (b) Carbohydrates; (c) Nutrients; (d) Minerals; (e) Essential amino acids

3. Under normal circumstances, protein function includes all of these *except:* (a) bone structure;
   (b) antibody formation; (c) ion pumps; (d) skeletal muscle structure; (e) energy reserves.

4. The ovaries and testes require cholesterol to synthesize their hormones. The cholesterol is
   delivered to them by: (a) chylomicrons; (b) HDLs; (c) LDLs; (d) VLDLs; (e) micelles.

5. Functions of the liver include all *except:* (a) transamination; (b) plasma protein synthesis;
   (c) detoxification of antibiotics; (d) manufacture of HDL shells; (e) production of hormones for
   carbohydrate metabolism.

6. Amino acids and fatty acids enter the citric acid cycle by way of: (a) pyruvic acid; (b) glycerol;
   (c) acetyl-CoA; (d) oxidative phosphorylation; (e) chemiosmosis.

7. Glucose-6-phosphate: (a) is used in maintaining favorable glucose concentration gradients across
   plasma membranes; (b) can be converted into fat; (c) can be converted to some amino acids;
   (d) can be utilized for ATP production; (e) all the above.

8. Urea, the major nitrogenous waste in urine, is produced by deamination of: (a) glutamic acid;
   (b) keto acids; (c) fatty acids; (d) monosaccharides; (e) dextrans.

9. When the air temperature is 38°C (100°F), the only way the body can lose heat is by:
   (a) thermoregulation; (b) hyperthermia; (c) convection; (d) radiation; (e) evaporation.

10. In the postabsorptive state: (a) the liver synthesizes glycogen; (b) the chylomicron level of the
    blood plasma is high; (c) the glucagon level is high; (d) gluconeogenesis is inhibited;
    (e) the insulin level is high.

11. The transfer of an –NH_2 group from an amino acid to α-ketoglutaric acid is: (a) transamination;
    (b) ketogenesis; (c) ketosis; (d) deamination; (e) amination.

12. Beta-oxidation is a process that: (a) removes glycerol from triglycerides; (b) polymerizes C_2
    groups to form fatty acids; (c) links fatty acids to glycerol; (d) breaks fatty acids into C_2
    fragments; (e) generates CO_2 in the citric acid cycle.

13. The conversion of amino acids to glucose is: (a) glycogenolysis; (b) glycolysis; (c) glucogenesis;
    (d) glycogenesis; (e) gluconeogenesis.

14. Most ATP is produced by: (a) glycolysis; (b) anaerobic fermentation; (c) beta-oxidation;
    (d) the mitochondrial matrix reactions; (e) the mitochondrial membrane reactions.

15. The energy to produce most ATP is transferred to the mitochondrial enzyme complexes by:
    (a) glucose; (b) pyruvic acid; (c) NADH; (d) $FADH_2$; (e) oxygen.

16. Which of these is a macronutrient? (a) vitamin C; (b) fat; (c) folic acid; (d) calcium; (e) sodium

17. Which of these is an essential nutrient? (a) starch; (b) protein; (c) sucrose; (d) linoleic acid; (e) all of these

18. The products of glycolysis include all *except:* (a) 2 ATP; (b) 2 pyruvic acid molecules; (c) 2 acetyl-CoA; (d) 2 H$^+$; (e) 2 NADH.

19. The role of hexokinase in glycolysis is to: (a) synthesize ATP; (b) split glucose into two 3-carbon sugars; (c) convert pyruvic acid to lactic acid; (d) transfer phosphate groups to NADH; (e) phosphorylate glucose.

20. The main reason for converting pyruvic acid to lactic acid in anaerobic fermentation is to: (a) regenerate NAD$^+$; (b) dispose of pyruvic acid, which is toxic to the cell; (c) prepare the pyruvic acid to enter the citric acid cycle; (d) hydrolyze ATP into ADP and P$_i$; (e) generate oxygen.

21. Which of these is/are true about metabolic processes?
    1. Glycogenolysis occurs more in the postabsorptive state than just after a meal.
    2. Glycogenesis occurs under the influence of many hormones, mainly epinephrine.
    3. Gluconeogenesis occurs when blood glucose levels are low, especially during a long fast.
    4. Glucagon is the process that releases ATP from glycogen breakdown.

    (a) 1 & 3; (b) 2 & 4; (c) 1, 2, & 3; (d) 4 only; (e) all the above

22. Which of these is/are true?
    1. In the liver, lactic acid can be converted into either glucose or glycogen.
    2. Coenzyme A is a B-vitamin derivative.
    3. Most of the CO$_2$ that we exhale comes from the citric acid cycle in the mitochondrial matrix.
    4. Of the original potential energy stored in a glucose molecule, most is transferred to the reduced coenzymes NADH and FADH$_2$.

    (a) 1 & 3; (b) 2 & 4; (c) 1, 2, & 3; (d) 4 only; (e) all the above

23. The absorptive state is characterized by:
    1. decreasing blood glucose levels.
    2. decreased gluconeogenesis.
    3. increased lipolysis.
    4. increased protein synthesis.

    (a) 1 & 3; (b) 2 & 4; (c) 1, 2, & 3; (d) 4 only; (e) all the above.

24. Which of these is/are true?
    1. Most Americans ingest more sugar and fat than RDAs suggest.
    2. Proteins are considered macronutrients because they are large molecules.
    3. Saturated fats can increase levels of blood cholesterol.
    4. Soluble fiber in the diet increases LDL levels.

    (a) 1 & 3; (b) 2 & 4; (c) 1, 2, & 3; (d) 4 only; (e) all the above

25. Which of these is/are true?
    1. Ghrelin stimulates the appetite.
    2. Obesity often occurs when infants and children are overfed.
    3. Leptin is a hormone that decreases appetite.
    4. Most obesity is caused by "bad" genes.

    (a) 1 & 3; (b) 2 & 4; (c) 1, 2, & 3; (d) 4 only; (e) all the above

E. Word Origins

1.   In *anorexia, -orexia* means "indigestion."
2.   In *hyperphagia, hyper-* means "excessive."
3.   In *hypoglycemia, glyc-* means "starch."
4.   In *glycolysis, -lysis* means "synthesis."
5.   In *cytochrome, -chrom* means "color."
6.   In *chemiosmotic, osmo-* means "diffusion."
7.   In *gluconeogenesis, neo-* means "new."
8.   In *thermoregulation, thermo-* means "heat."
9.   In *lipolysis, lipo-* means "fat."
10. In *kilocalorie, kilo-* means "thousand."
11. asc-
12. lept-
13. -ites
14. an-
15. -emesis

F. Which One Does Not Belong?

1. (a) citric acid; (b) phosphoglyceraldehyde; (c) oxaloacetic acid; (d) acetyl-CoA

2. (a) retinol; (b) riboflavin; (c) phylloquinone; (d) calcitriol

3. (a) matrix reactions; (b) membrane reactions; (c) anaerobic fermentation; (d) glycolysis

4. (a) NADH; (b) cytochrome c; (c) coenzyme Q; (d) flavin dinucleotide

5. (a) vitamin B_2; (b) vitamin A; (c) vitamin E; (d) vitamin C

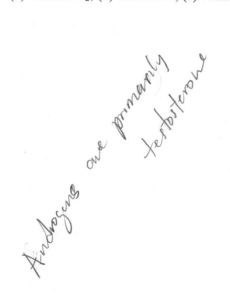

Androgens are primarily testosterone

27   The Male Reproductive System

A. Short Answer

1. If an embryo has a Y chromosome, its gonads respond to an SRY gene product called ___.

2. The labia majora of a female and the scrotum of a male are ___ because they both arise from the same structure, the labioscrotal folds of the embryo.

3. The descending fetal testes are guided through the inguinal canal into the ___ by a cord called the ___.

4. The external genitalia and anal orifice occupy a diamond-shaped space between the thighs called the ___.

5. Sperm are produced in tiny ducts in the testes called the ___.

6. In mammals, the ___ determines the gender of the offspring.

7. The spermatic cord contains the ___ plexus of blood vessels, which behaves as a(n) ___ exchanger to prevent the incoming blood from overheating the testes.

8. A short passage, the _Ejaculatory duct_, conveys both sperm and seminal vesicle fluid to the prostatic urethra.

9. In a male embryo, _Testosterone_ stimulates the descent of the testes into the scrotum.

10. Erection of the penis results from the swelling of tiny cavities, called _Lacunae_, with blood.

11. In a female, LH causes ovulation; in a male, it stimulates the interstitial cells to secrete _Androgen / Testosterone_

12. The hormone _Testosterone_ is responsible for stimulating the sex drive, or _Libido_, in both sexes.

13. The chromosomes we pass on to our children are not the same as those we inherit from our parents because they exchange DNA with each other during a process called ___ in meiosis. _Crossing over_

14. _Chryptorchidism_ ___ is a condition in which baby boys are born with undescended testes.

15. Semen is expelled from the penis by rhythmic contractions of the _bulbospongiosus_ ___ muscle that ensheaths its root.

B. Matching

A. midpiece	G. germ cells	M. ovum	S. sustentacular cells
B. testes	H. anaphase II	N. inhibin	T. bulbospongiosus
C. seminal vesicles	I. dartos	O. cremaster	U. meiosis II
D. FSH	J. testosterone	P. sperm	V. mesonephric ducts
E. SRY	K. interstitial cells	Q. prostate	W. paramesonephric ducts
F. ABP	L. zygote	R. frenulum	X. urogenital folds

1. Embryonic forerunner of the male reproductive tract *mesonephric ducts*
2. Protein from sustentacular cells that binds testosterone *ABP*
3. Stratified cells of the seminiferous tubule *Germ Cell*
4. Cells that form the blood–testis barrier
5. Diploid product of the union of gametes
6. Muscle of the spermatic cord
7. Produce(s) most of the semen
8. Slows down sperm production without affecting testosterone secretion *Inhibin*
9. Stage in which each chromosome divides into two independent chromatids
10. Location of the mitochondria in a spermatozoon

C. True or False

1. In *Homo sapiens,* a haploid cell has 23 pairs of chromosomes, and a diploid cell has 46 pairs.

2. Meiosis II converts a germ cell from a diploid to a haploid state.

3. Homologous chromosomes are physically but not genetically identical.

4. During puberty, testosterone stimulates an increase in EPO, while DTH stimulates body hair growth.

5. The conversion of spermatids to spermatozoa involves mitosis, not meiosis.

6. Drugs used to treat erectile dysfunction act by increasing blood flow to the penis.

7. In the last several decades, testicular cancer has tripled, and sperm counts have dropped by nearly 50%.

8. Sexual reproduction requires combining genetic material from two individuals.

9. Semen consists mostly of spermatozoa.

10. Both ovaries and testes descend from their early embryonic positions in the abdominal cavity.

D. Multiple Choice

1. The male has two of all the following *except:* (a) bulbourethral gland(s); (b) corpus cavernosum (corpora cavernosa); (c) prostate(s); (d) seminal vesicle(s); (e) crus (crura) of the penis.

2. All of these are true about the prostate gland *except:* (a) in most men, it continues to grow in size after middle age; (b) it surrounds the ejaculatory duct; (c) the PSA test can aid in diagnosing prostate cancer; (d) its glandular secretions make up about 30% of semen; (e) it secretes large amounts of fructose for sperm nourishment.

biological Half life
anion -
cation +

Lanugo
Vellus

3. The smooth muscle in the wall of the scrotum is the: (a) cremaster; (b) muscularis interna; (c) tunica media; (d) tunica dartos; (e) tunica vaginalis.

4. The principle site of sperm storage until ejaculation is the: (a) penis; (b) epididymis; (c) seminal vesicles; (d) prostate; (e) seminiferous tubules.

5. The formation of spermatozoa from haploid spermatids is called: (a) spermiogenesis; (b) spermatogenesis; (c) primary spermatogenesis; (d) meiosis; (e) spermatogonia.

6. The most common bacterial STD in the United States is: (a) HIV; (b) herpes; (c) HCV; (d) chlamydia; (e) syphilis.

7. The hormone that initiates the changes of puberty is (a) gonadotropin-releasing hormone; (b) interstitial cell–stimulating hormone; (c) follicle-stimulating hormone; (d) inhibin; (e) testosterone.

8. A male mammal receives his Y chromosome from his: (a) mother; (b) grandmother; (c) father; (d) offspring; (e) fraternal grandfather.

9. Sperm penetrate eggs by means of enzymes in their: (a) midpiece; (b) acrosome; (c) tail; (d) nucleus; (e) principal piece.

10. During puberty, testosterone: (a) stimulates ABP synthesis; (b) stimulates libido; (c) stimulates production of inhibin in sustentacular cells; (d) inhibits erythropoiesis; (e) stimulates axillary and pubic hair growth.

11. All of these are true about the male climacteric *except*: (a) LH and FSH levels increase; (b) feedback inhibition of the pituitary decreases; (c) it renders men unable to father children; (d) it can cause hot flashes and mood changes; (e) testosterone levels drop.

12. The peristaltic propulsion of sperm from the epididymis to the ampulla of the ductus deferens is: (a) emission; (b) detumescence; (c) myotonia; (d) vasocongestion; (e) intromission.

13. Fibrinogen is produced not only by the liver for blood clotting but also by the ___ for semen clotting. (a) ejaculatory ducts; (b) testes; (c) seminiferous tubules; (d) prostate; (e) seminal vesicles

14. Functions of the sustentacular cells include all *except:* (a) promoting sperm formation; (b) secreting inhibin (c) maintaining the blood–testis barrier; (d) connecting to the rete testis; (e) all are functions of these cells.

15. In humans, one spermatozoon contains ___ chromosomes. (a) 46; (b) 23; (c) 10; (d) 4; (e) 2

16. The function of müllerian-inhibiting factor is to: (a) suppress the development of a male reproductive tract; (b) suppress the development of a female reproductive tract; (c) stimulate differentiation of male external genitalia; (d) stimulate differentiation of female external genitalia; (e) activate the secretion of gonadotropins.

17. All of these are secondary sex characteristics *except:* (a) the penis; (b) a deep voice; (c) pubic hair; (d) larger muscle mass; (e) facial hair.

18. Hypospadias is an anatomical abnormality of the: (a) urethra; (b) testes; (c) prostate; (d) pampiniform plexus; (e) scrotum.

19. Which of the following plays a role in thermoregulation of the testes? (a) the tunica vaginalis; (b) the tunica albuginea; (c) the dartos; (d) the erectile tissues; (e) the epididymis

20. When a vasectomy is performed, the ___ is cut. (a) ductus deferens; (b) urethra; (c) ejaculatory duct; (d) epididymis; (e) seminiferous tubule

21. Which of these is/are true about STDs?
    1. An STD can be contagious even if no symptoms are noticeable.
    2. The virus that causes genital herpes can move along nerve fibers to other parts of the body.
    3. PID often causes sterility in women.
    4. There are no vaccines for prevention of most STDs.

    (a) 1 & 3; (b) 2 & 4; (c) 1, 2, & 3; (d) 4 only; (e) all the above

22. Which of these is/are true?
    1. Testicular feminization (AIS) occurs when inadequate amounts of testosterone are produced by a male.
    2. In warm weather, the cremaster muscle is contracted, so the testes are farther from the body.
    3. The pampiniform plexus is derived from the abdominal aorta.
    4. An XY male genotype can result in a female phenotype.

    (a) 1 & 3; (b) 2 & 4; (c) 1, 2, & 3; (d) 4 only; (e) all the above

23. Which of these is/are true?
    1. By the end of the third month of development, genitalia in both sexes are formed.
    2. The vas deferens, spermatic artery, and vein and lymphatic vessels all pass through the inguinal canal.
    3. Development of a female embryo occurs in the absence of androgens.
    4. Primordial germ cells from the yolk sac become spermatogonia early in embryonic development.

    (a) 1 & 3; (b) 2 & 4; (c) 1, 2, & 3; (d) 4 only; (e) all the above

24. Which of these is/are true?
    1. Meiosis maintains the normal diploid chromosome number from generation to generation.
    2. Crossing over occurs in prophase I, when homologous chromosomes come together to form a tetrad.
    3. Both divisions of the ANS are involved in the sexual response in males.
    4. Genetic variation produced by sexual reproduction is the basis for evolution of a species.

    (a) 1 & 3; (b) 2 & 4; (c) 1, 2, & 3; (d) 4 only; (e) all the above

25. In the hormonal feedback mechanism in the male reproductive system:
    1. GnRH from the anterior pituitary stimulates the interstitial cells.
    2. follicle-stimulating hormone triggers the production of androgen-binding protein in sustentacular cells.
    3. inhibin temporarily prevents the production of sperm.
    4. luteinizing hormone stimulates the production of androgens.

    (a) 1 & 3; (b) 2 & 4; (c) 1, 2, & 3; (d) 4 only; (e) all the above

E. Word Origins

1. In *gamete, gam-* means "marriage."
2. In *cryptorchidism, crypt-* means "tomb."
3. In *cryptorchidism, orchid-* means "flower."
4. In *tunica vaginalis, vagina-* means "female."
5. In *seminiferous, semin-* means "seed, sperm."
6. In *perineal raphe, raphe* means "wall, partition."
7. In *epididymis, didym-* means "twin, testis."
8. In *prostate, -stat* means "stand."
9. In *acrosome, acro-* means "sharp, pungent."
10. In *ejaculate, jacul-* means "throw."
11. helic-
12. gubern-
13. herp-
14. rete
15. puber-

F. Which One Does Not Belong?

1. (a) sperm; (b) spermatogonia; (c) spermatids; (d) secondary spermatocyte

2. (a) prostate; (b) penis; (c) vas deferens; (d) testes

3. (a) gonorrhea; (b) genital warts; (c) syphilis; (d) chlamydia

4. (a) seminiferous tubules; (b) sustentacular cells; (c) Leydig cells; (d) gubernaculum

5. (a) stimulates spermatogenesis; (b) causes growth spurt at puberty; (c) increases GnRH from hypothalamus; (d) stimulates erythropoiesis

G. Figure Exercise

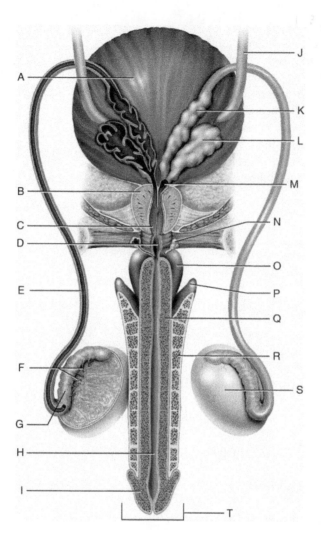

Match the statements with the structures in the diagram. Some answers may be used more than once; some questions can have more than one answer.

1.   Erectile body on ventral penis
2.   Tube(s) mature sperm must pass through to leave body
3.   Secrete(s) fructose
4.   Cut in sterilization surgery
5.   Contain lacunae
6.   Store(s) sperm for about 20 days
7.   Endocrine gland
8.   Cannot pass semen and urine at the same time
9.   Required for synthesis of seminogelin
10. Target for FSH and LH

28   The Female Reproductive System

A. Short Answer

1. Each egg develops in a vesicle of the ovary, called a(n) ___, and is released in midcycle by the rupture of that vesicle in a process called ___.

2. After the ovary releases an egg, the ciliated mucosa of the ___ conveys it to the uterus.

3. Childbirth is achieved partly by contractions of the thick, muscular wall of the uterus, called the ___.

4. Menstrual fluid is composed of blood and the surface layer, or stratum functionalis, of the ___.

5. All the external genitalia of a female are collectively called the ___.

6. The female structure homologous to the male's glans penis is the ___.

7. Each milk duct of the mammary gland ends with a dilation called the ___ just before it opens on the nipple.

8. The flow of milk from the mammary acini down their ducts is induced by the contraction of ___ cells under the influence of the hormone ___.

9. The midlife cessation of the menses is called ___, and it occurs during a transitional period of life called the ___.

10. Androgens of the adrenal cortex and ovary stimulate ___, the growth of the pubic and axillary hair at puberty.

11. In the last stage of oogenesis before meiosis I, the egg is called the ___, and the cell that is ovulated is the ___.

12. During fetal development in a female, the number of primary oocytes dramatically ___ due to ___.

13. When an egg is ovulated, it is surrounded by a gelatinous layer called the ___ and, outside of that, a layer of cells called the ___.

14. In the second half of the menstrual cycle, progesterone from the ___ stimulates the endometrial cells of the uterus to produce glycogen and mucus.

15. During pregnancy, the pituitary hormone ___ is secreted, but it has no effect on the mammary glands. After birth, it stimulates milk synthesis.

B. Matching

A. puerperium	G. premenstrual phase	M. relaxin	S. lochia
B. pubarche	H. granulosa cells	N. oogonium	T. secondary oocyte
C. menarche	I. involution	O. oxytocin	U. proliferative phase
D. secretory phase	J. primary oocyte	P. estrogen	V. myoepithelial cells
E. ampullae	K. corpus luteum	Q. corona radiata	W. uterosacral ligament
F. fimbriae	L. progesterone	R. broad ligament	X. suspensory ligament

  1. Sheet of peritoneum on each side of the uterus
  2. Feathery projections of the distal end of the uterine tube
  3. A girl's first menstrual period
  4. Promotes development of blood vessels in the pregnant uterus
  5. Peaks during secretory phase of menstrual cycle
  6. Stratified epithelial cells that line a mature ovarian follicle
  7. Period in which the endometrium thickens by mitosis
  8. Progesterone-secreting body that develops from an ovulated follicle
  9. Hormone central to the positive feedback theory of labor
10. Shrinkage of the postpartum uterus

C. True or False

1. After ovulation, a follicle begins to move down the uterine tube to the uterus.

2. An ovum never completes meiosis II unless it is fertilized.

3. Modern pregnancy tests are based on the detection of HCG in the urine.

4. HCG is secreted by the granulosa cells of the ovarian follicle.

5. The vestibular bulbs are erectile tissues of the clitoris.

6. Cow's milk does not contain enough protein for adequate infant nutrition.

7. Decidual cells synthesize glycogen during the secretory phase of the menstrual cycle.

8. Lochia results from the postpartum autolysis of the uterus.

9. New follicles do not begin to develop until the menstrual period is over.

10. The dominant hormone of the follicular phase of the ovarian cycle is estrogen.

D. Multiple Choice

1. A Pap smear is a means of detecting: (a) breast cancer; (b) cervical cancer; (c) genital herpes; (d) uterine cancer; (e) pregnancy.

2. The first germ cells of the female develop in: (a) puberty; (b) the primordial follicles; (c) the germinal epithelium; (d) the yolk sac during embryonic development; (e) the ovarian fossa.

3. Effacement is a change in the tissue of the: (a) endometrium; (b) ovary; (c) cervix; (d) uterine tube; (e) mammary gland.

4. ___ begins at menarche, but ___ does not begin until about a year later. (a) Menstruation, ovulation; (b) Breast development, pubic hair growth; (c) Puberty, bone growth; (d) Puberty, adolescence; (e) Ovulation, oogenesis

5. The hormone that peaks during the menstrual phase of the menstrual cycle is: (a) luteinizing hormone; (b) progesterone; (c) estrogen; (d) FSH; (e) none of these.

6. LH secretion peaks: (a) just prior to ovulation; (b) at the end of the luteal phase; (c) at ovulation; (d) at the same time progesterone secretion peaks; (e) during the menstrual phase.

7. Luteinizing hormone triggers: (a) micturition; (b) parturition; (c) gestation; (d) lactation; (e) ovulation.

8. The proliferative phase of the menstrual cycle occurs during the ___ phase of the ovarian cycle. (a) ischemic; (b) climacteric; (c) follicular; (d) luteal; (e) postovulatory

9. As pregnancy progresses, the hormonal role of the corpus luteum is taken over by: (a) the placenta; (b) the anterior pituitary; (c) the posterior pituitary; (d) the fetus; (e) secondary follicles of the ovary.

10. Childbirth is also called: (a) gestation; (b) emission; (c) expulsion; (d) effacement; (e) parturition.

11. Labor contractions are stimulated partly by a placental secretion called: (a) luteinizing hormone; (b) prostaglandin; (c) HCG; (d) oxytocin; (e) meconium.

12. The mammary glands secrete ___, not milk, for the first couple of days after childbirth. (a) meconium; (b) lochia; (c) lanugo; (d) colostrum; (e) vernix

13. Suckling by a baby stimulates the secretion of: (a) estrogen; (b) progesterone; (c) human placental lactogen; (d) prolactin; (e) luteinizing hormone.

14. Breast size is determined by: (a) the number of lactiferous sinuses; (b) adipose tissue; (c) the number of lobes; (d) the number of lobules; (e) hormones.

15. Milk synthesis requires all *except:* (a) cortisol; (b) insulin; (c) growth hormone; (d) parathyroid hormone; (e) human chorionic somatomammotropin.

16. The female reproductive system develops : (a) from müllerian ducts ; (b) under the influence of gonadotropins; (c) without any hormone intervention; (d) in the beginning stages of pubarche; (e) in the same basic ways that male systems develop.

17. The function of the vestibular glands of the female is to: (a) produce eggs; (b) hold milk until the infant sucks it out; (c) constrict the orgasmic platform; (d) secrete the lubricant needed for intercourse; (e) produce vasocongestion of the clitoris.

18. The beginning of breast development is called: (a) pubarche; (b) climacteric; (c) menarche; (d) thelarche; (e) atresia.

19. Lochia is: (a) secreted by the CL; (b) a product of involution and autolysis; (c) an unreliable form of birth control; (d) from the ovary and produces progesterone; (e) amplified just before menopause, then drops off sharply.

20. During pregnancy, the ducts of the mammary glands grow and branch under the influence of:
(a) estrogen; (b) progesterone; (c) human placental lactogen; (d) prolactin; (e) oxytocin.

21. Menopausal changes include:
    1. an increased number of "eggless" follicles.
    2. atrophy of the reproductive system.
    3. decreased cholesterol levels.
    4. decreased gonadotropin effects on ovarian follicles.

    (a) 1 & 3; (b) 2 & 4; (c) 1, 2, & 3; (d) 4 only; (e) all the above

22. Which of these is/are true about contraception?
    1. Barrier methods include the condom, diaphragm, and sponge.
    2. Tubal ligation and vasectomy are forms of sterilization and are the most effective forms of contraception.
    3. Hormones in the birth control pill, skin patches, and vaginal rings prevent ovulation and, therefore, conception.
    4. IUDs prevent implantation.

    (a) 1 & 3; (b) 2 & 4; (c) 1, 2, & 3; (d) 4 only; (e) all the above

23. Which of these is/are true about hormonal changes during pregnancy?
    1. Estrogen levels are higher at the end of gestation than in the beginning, but HCG shows the opposite pattern.
    2. HCS, produced by the placenta, mobilizes fatty acids for fetal fuel.
    3. Progesterone and estrogen prevent ovulation by inhibiting anterior pituitary secretion of FSH and LH.
    4. Pituitary thyrotropin stimulates the initial stages of parturition.

    (a) 1 & 3; (b) 2 & 4; (c) 1, 2, & 3; (d) 4 only; (e) all the above

24. Which of these is/are true about the female sexual cycle?
    1. The optimal fertile period is around the time of ovulation.
    2. Endometrial ischemia and necrosis occur during the premenstrual phase.
    3. During the secretory phase, the corpus luteum forms and secretes progesterone.
    4. Ovulation requires a positive feedback mechanism involving LH, FSH, and GnRH.

    (a) 1 & 3; (b) 2 & 4; (c) 1, 2, & 3; (d) 4 only; (e) all the above

25. Which of these is/are true?
    1. The second polar body is not produced if fertilization does not occur.
    2. The number of primary oocytes increases as puberty approaches.
    3. Meiosis I produces one polar body and one secondary oocyte.
    4. Oogenesis, like spermatogenesis, is cyclic in nature.

    (a) 1 & 3; (b) 2 & 4; (c) 1, 2, & 3; (d) 4 only; (e) all the above

E.   Word Origins

1.   In *ovary, -ary* means "place for."
2.   In *hysterectomy, hyster-* means "ovary."
3.   In *hysterectomy, -ectomy* means "cutting out."
4.   In *endometrium, -metr* means "measure."
5.   In *lactiferous, lacti-* means "milk."
6.   In *estrogen, estro-* means "female."
7.   In *menstruation, men-* means "bloody."
8.   In *progesterone, pro-* means "favoring, promoting."
9.   In *oogenesis, oo-* means "proliferate."
10.   In *hyperemesis, -emesis* means "vomiting."
11.   neo-
12.   gest-
13.   puer-
14.   -schmerz
15.   lute-

F.   Which One Does Not Belong?

1.   (a) decreased P_{CO_2}; (b) increased blood volume; (c) increased GFR; (d) decreased ventilation rate

2.   (a) progesterone; (b) GnRH; (c) estrogen; (d) inhibin

3.   (a) increased progesterone; (b) increased LH; (c) increased collagenase; (d) increased body temperature

4.   (a) mifepristone; (b) IUD; (c) diaphragm; (d) Levonelle

5.   (a) increased ACTH; (b) decreased aldosterone; (c) increased relaxin; (d) decreased HCG

G.   Figure Exercise

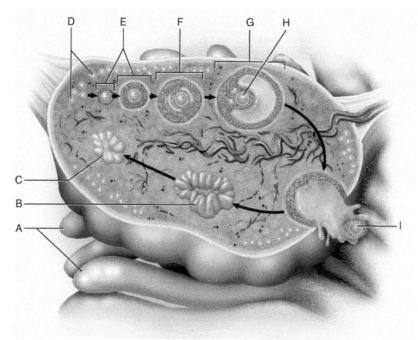

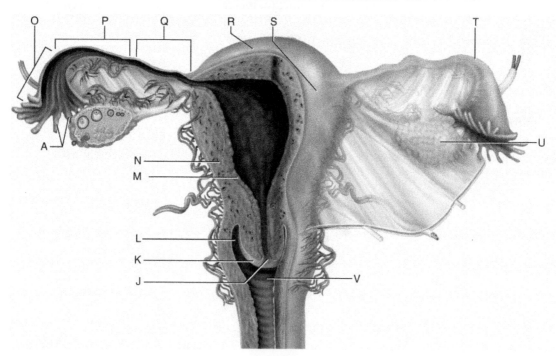

Match the statements with the structures in the diagram. Some answers may be used more than once; some questions can have more than one answer.

1. In the adult ovary, over 90% of follicles are this type
2. Enclosed in the mesosalpinx
3. Place for fertilization if normal implantation is to occur
4. Undergoes metaplasia from childhood to puberty
5. Lined with corrugated mucosa and ciliated cells
6. Their cilia enhance movement of oocyte into uterine tube
7. Has a mucous plug that changes in consistency with the phases of the sexual cycle
8. First to contain granulosa cells that secrete the zona pellucida around the oocyte
9. Contain(s) smooth muscle
10. Can be recognized by increased body temperature and/or *mittelschmerz*
11. Reaches peak growth in the secretory phase of menstrual cycle
12. Secretes progesterone in the second half of ovarian cycle
13. Epithelial cells ferment glycogen to lactic acid, which lowers pH
14. Has OT receptors late in pregnancy
15. Cells secrete inhibin

29   Human Development and Aging

A. Short Answer

1. Sperm cannot fertilize an egg until they have undergone the process of ___.

2. The ___ of an egg undergo exocytosis to produce a slow block to polyspermy.

3. Our sense of hearing peaks in ___.

4. ___ twins are the result of the union of one egg and one sperm.

5. During pregnancy, the ___ becomes a major endocrine gland.

6. The failure of two homologous chromosomes to separate from each other at anaphase I is called ___.

7. The greatest age attainable by human beings is called the ___.

8. Some of the senescent changes of the skin, called ___, are due to a person's lifelong exposure to sunlight.

9. Chorionic villi from the ___ digest their way into the endometrium toward uterine blood vessels.

10. Nondisjunction of the 21st chromosome causes ___.

11. The fetus is suspended in ___, composed in large part of its own urine.

12. In the seventh month of development, the fetus turns into a head-down ___ position.

13. Prior to birth, a shunt called the ___ between the ascending aorta and pulmonary trunk allows most blood to bypass the lungs.

14. The single diploid cell formed from the fusion of egg and sperm nuclei is called the ___.

15. A premature neonate cannot regulate its own body temperature because its ___ is not well developed.

B. Matching

A. cleavage	G. alcohol	M. polyspermy	S. Alzheimer disease
B. embryoblast	H. osteopenia	N. ductus venosus	T. Parkinson disease
C. foramen ovale	I. Down syndrome	O. glaucoma	U. ductus arteriosus
D. uterine milk	J. meconium	P. X rays	V. thalidomide
E. Turner syndrome	K. decidual cells	Q. meiosis	W. amniotic fluid
F. osteoarthritis	L. cataracts	R. vernix caseosa	X. trophoblast

1. The inner cell mass of a blastocyst
2. Prevented by the rapid influx of Na^+ into the egg after sperm penetration
3. Cloudiness of the lenses common in old age
4. Teratogen that causes limb deformities
5. Characterized by neurofibrillary tangles and amyloid plaques
6. Loss of bone tissue
7. Condition in which females have only one X chromosome
8. Division resulting in increasing numbers of smaller blastomeres
9. Cheesy secretion on the fetal skin
10. Circulatory shunt that bypasses the fetal liver

C. True or False

1. An oocyte must be fertilized within 72 hours after ovulation or it will die.

2. The first sperm to reach an egg is not likely to be the one that fertilizes it.

3. The yolk sac serves no purpose in humans; it is a vestigial organ.

4. The most common cause of infant mortality is birth defects.

5. Alcohol is the most common cause of birth defects in the United States.

6. Aneuploidy is the result of nondisjunction.

7. Trophoblastic nutrition continues for a time after placental nutrition begins.

8. Elderly people are at increased risk of heatstroke because they have more subcutaneous fat, which retains body heat.

9. Fertilization usually occurs in the uterus.

10. Scientists are optimistic that, with genetic engineering and the conquest of infectious diseases, the human life span will soon be extended to 150 years or more.

D. Multiple Choice

1. Which of the following is *not* one of the embryonic membranes? (a) amnion; (b) ectoderm; (c) allantois; (d) yolk sac; (e) chorion

2. The outer cells of a blastocyst, responsible for implantation and early nutrition of the conceptus, are called the: (a) ectoderm; (b) trophoblast; (c) embryophore; (d) fertilization membrane; (e) blastomeres.

3. At the time a conceptus arrives in the uterus, it is in the ___ stage of development. (a) morula; (b) zygote; (c) neonatal; (d) embryonic; (e) blastocyst

4. A neonate's first bowel movement expels ___ from the large intestine. (a) chyme; (b) lanugo; (c) meconium; (d) vernix caseosa; (e) colostrums

5. Which of these provides a way for venous blood to bypass the lungs of a fetus? (a) the fossa ovalis; (b) the urachus; (c) the pulmonary shunt; (d) the foramen ovale; (e) the ductus venosus

6. The criterion for calling a developing individual an embryo is the presence of: (a) embryonic membranes; (b) amniotic fluid; (c) a primitive streak; (d) primary germ layers; (e) distinguishable organ systems.

7. Prior to implantation, a conceptus is nourished by: (a) amniotic fluid; (b) the trophoblastic mode; (c) colostrum; (d) maternal blood; (e) uterine milk.

8. All of these are true about premature infants *except:* (a) they suffer from congenital abnormalities; (b) they have hypoproteinemia; (c) they have a poorly developed swallowing reflex; (d) they often die of IRDS; (e) they are unable to synthesize vitamin K.

9. Which of the following membranes contributes to the structure of the placenta? (a) allantois; (b) chorion; (c) yolk sac; (d) amnion; (e) mesoderm

10. Agents that cause fetal abnormalities are called: (a) mutagens; (b) antigens; (c) teratogens; (d) estrogens; (e) telogens.

11. Nondisjunction is *not* responsible for: (a) Down syndrome; (b) aneuploidy; (c) Turner syndrome; (d) hydrocephalus; (e) Klinefelter syndrome.

12. Implantation of a blastocyst outside the uterus: (a) is common in mothers younger than 21; (b) often results in a healthy baby; (c) can be reversed with surgery; (d) is more common in women in their 40s; (e) is called an ectopic pregnancy and can be life-threatening to the mother.

13. The COPDs are degenerative diseases of the ___ and a major cause of death in old age. (a) lungs; (b) kidneys; (c) nervous system; (d) cardiovascular system; (e) skeletal system

14. The ___ secretes HCG. (a) blastocyst; (b) trophoblast; (c) syncytiotrophoblast; (d) cytotrophoblast; (e) embryo

15. The mesoderm gives rise to all of the following *except:* (a) muscles; (b) the brain; (c) blood vessels; (d) bones; (e) ovaries.

16. The function of the acrosomal reaction is to: (a) clear a path to the egg; (b) prevent polyspermy; (c) produce an acrosome; (d) capacitate the sperm; (e) initiate cleavage.

17. In human development, what stage comes between the zygote and the blastocyst? (a) the oocyte; (b) the neonate; (c) the morula; (d) the embryoblast; (e) the embryo

18. In trophoblastic nutrition, the source of nutrients for the conceptus is: (a) colostrum; (b) the decidual cells; (c) the corpus luteum; (d) uterine milk; (e) the mother's bloodstream.

19. One theory of aging relates senescence to a shortening of the ___ of each chromosome every time a cell divides. (a) telomeres; (b) telophases; (c) centrioles; (d) centromeres; (e) spindles

20. Which of these is correct about meiosis? (a) It produces four genetically identical gametes in males but not in females. (b) In males, it begins in embryonic germ cells. (c) In females, it begins in menarche and ends in menopause. (d) In females, it can take as long as 50 years. (e) In both sexes, it ends in late middle age.

21. Which of these is/are true?
    1. Genes that do not alter the rate of reproduction are not subject to natural selection.
    2. IVF can result in multiple births.
    3. It is possible for a woman to give birth to her own grandchildren, nieces, and nephews.
    4. Advances in medical techniques are often ahead of the ethical and legal ramifications they create.

    (a) 1 & 3; (b) 2 & 4; (c) 1, 2, & 3; (d) 4 only; (e) all the above

22. Which of these is/are true about aging?
    1. Senescence is synonymous with aging.
    2. The most common nervous system disorder in the elderly is Alzheimer disease.
    3. The number one cause of death in the elderly is cancer.
    4. Constipation is the most universal complaint of the elderly.

    (a) 1 & 3; (b) 2 & 4; (c) 1, 2, & 3; (d) 4 only; (e) all the above

23. Which of these is/are true?
    1. Placental sinuses filled with maternal blood surround chorionic villi that contain fetal blood, but the two do not mix.
    2. Brown fat in babies produces heat rather than ATP.
    3. About 50% of spontaneous abortions are caused by aneuploidy.
    4. The placenta is permeable to harmful chemicals in maternal blood.

    (a) 1 & 3; (b) 2 & 4; (c) 1, 2, & 3; (d) 4 only; (e) all the above

24. Which of these is/are true about senescence?
    1. Free radical damage to macromolecules may be a contributing factor.
    2. Collagen molecules become cross-linked with disulfide bridges.
    3. It is too often a result of obesity and lack of exercise.
    4. Levels of telomerase increase in aging cells, making them incapable of dividing.

    (a) 1 & 3; (b) 2 & 4; (c) 1, 2, & 3; (d) 4 only; (e) all the above

25. Which of these is/are true?
    1. The conceptus spends about three days traveling from the distal uterine tube to the uterus, where it remains for another four to five days before implantation.
    2. At the time of fertilization, the sperm nucleus and midpiece enter the egg, but no paternal mitochondria survive.
    3. Progeria is a genetic condition in which senescence is greatly accelerated.
    4. Gene defects are the leading cause of death in newborns.

    (a) 1 & 3; (b) 2 & 4; (c) 1, 2, & 3; (d) 4 only; (e) all the above

E. Word Origins

1. In *embryoblast, -blast* means "container."
2. In *trophoblast, tropho-* means "nourishment."
3. In *ectopic, -ic* means "pertaining to."
4. In *corpus luteum, lute-* means "hormonal."
5. In *decidual cells, decid-* means "falling off."
6. In *neonate, -nate* means "baby."
7. In *teratogen, terato-* means "mutant."
8. In *aneuploidy, eu-* means "true, normal."
9. *Placenta-* means "flat cake."
10. In *telomere, telo-* means "determining."
11. ger-
12. con-
13. mor-
14. nom-
15. syn-

F. Which One Does Not Belong?

1. (a) speed of action potentials; (b) number of glomeruli; (c) number of cutaneous nerve endings; (d) number of fast-twitch muscle fibers

2. (a) umbilical cord; (b) yolk sac; (c) chorion; (d) amnion

3. (a) thalidomide; (b) gonorrhea; (c) alcohol; (d) smoking

4. (a) ductus venosus; (b) foramen ovale; (c) ductus arteriosus; (d) ligamentum venosum

5. (a) adrenal cortex; (b) blood; (c) ureters; (d) adrenal medulla

G. Identification Exercise

For the following items, indicate whether each increases or decreases with age.

1. Lymphatic tissue and bone marrow
2. Respiratory infections
3. Number of cutaneous nerve endings
4. Velocity of action potentials
5. Calcium deficiency
6. COPDs
7. Kidney size, number of glomeruli
8. Brain weight
9. GFR
10. Prostatic hyperplasia
11. Atherosclerosis
12. Heart muscle mass
13. Heartburn
14. Caloric intake
15. Thrombosis
16. Myofibrils
17. Alveoli
18. Reaction time
19. Proportion of body fat
20. Long-term memory

Part III
Answer Key

Practice Exams

Chapter 1: Major Themes of Anatomy and Physiology

A. Short Answer

1. gross anatomy
2. homeostasis
3. deduction
4. control group
5. peer review
6. theory
7. Charles Darwin
8. selection pressures
9. negative feedback
10. tissues
11. organelles
12. effector
13. excretion
14. hypothesis
15. opposable, prehensile

B. Matching

| 1. S | 3. G | 5. A | 7. M | 9. X |
| 2. C | 4. D | 6. E | 8. N | 10. B |

C. True or False

The following items are false for the reasons stated. The other five are true.
1. Vesalius and his students were allowed to dissect cadavers; this was taboo in Galen's time.
3. It is the tissue level. Organelles are smaller components of a cell.
4. A baroreflex is an example of a homeostatic mechanism that regulates blood pressure.
6. There are many processes, such as blood clotting and ovulation, in which positive feedback is normal and essential.
8. Harvey speculated that such connections must exist but was unable to observe them for lack of a microscope, an instrument developed after his time.

D. Multiple Choice

1. b	6. c	11. c	16. b	21. b
2. b	7. a	12. c	17. c	22. a
3. a	8. e	13. a	18. e	23. e
4. d	9. d	14. c	19. e	24. c
5. c	10. b	15. a	20. b	25. a

E. Word Origins and Meanings

1. No, cutting
2. Yes
3. Yes
4. No, within
5. Yes
6. Yes
7. No, tissue
8. No, tree
9. Yes
10. Yes
11. apart; dissection, disease
12. the same; homeostasis
13. nature; physiology
14. cut; dissection
15. to fail or die; cadaver

F. Which One Does Not Belong?

1. (a) is not a selection pressure, as are the others
2. (d) the rest allowed for bipedalism
3. (b) others are all theories
4. (c) named the cell; all others made contributions to the concept of homeostasis
5. (a) is plural; others are singular terms

Atlas A: A General Orientation to Human Anatomy

A. Short Answer

1. pronated
2. median (midsagittal)
3. superficial, deep
4. nose or forehead/tail
5. axillary, cubital
6. carpal, tarsal
7. umbilical, hypogastric
8. epigastric, hypochondriac
9. mediastinum, pleural
10. parietal, visceral
11. dorsum or dorsal surfaces
12. nervous, endocrine
13. circulatory, lymphatic
14. respiratory, urinary
15. distal, proximal

B. Matching

1. G	3. N	5. U	7. X	9. Q
2. B	4. C	6. K	8. L	10. W

C. True or False

The following items are false for the reasons stated. The other five are true.

1. It can be described as anterior or superficial to the heart but is not inferior to it in humans.
4. There is an indefinite number of parasagittal planes, but only one true median (midsagittal) plane.
5. A frontal section would show all four chambers, but a cross section would show only two.
6. Some organs belong to two or more organ systems. For example, the penis is part of the urinary and reproductive systems, and the pharynx is part of the digestive and respiratory systems.
7. They are in the thoracic cavity, which is divided into the pleural cavities (lungs) and pericardial cavities (heart).

D. Multiple Choice

1. b	6. c	11. a	16. a	21. c
2. a	7. b	12. b	17. a	22. d
3. b	8. e	13. e	18. e	23. a
4. c	9. a	14. c	19. d	24. b
5. b	10. a	15. b	20. c	25. b

E. Word Origins and Meanings

1. No, next to
2. No, rib
3. Yes
4. No, cartilage
5. No, before
6. Yes
7. No, side or rib
8. Yes
9. No, behind
10. No, intestine
11. above, over; epigastric
12. below; hypogastric
13. around; peritoneum
14. below; subcostal
15. stomach; epigastric

F. Which One Does Not Belong?

1. (d) is a body plane; the others are directional terms
2. (b) is the bottom of the foot; the rest are associated with the antebrachium
3. (c) consist of three layers; while the others have only two layers;
4. (a) is not a true system; others are input/output systems
5. (b) is in the peritoneal cavity; the others are retroperitoneal

Chapter 2: The Chemistry of Life

A. Short Answer

1. polymers
2. valence electrons
3. cation, anion
4. free radicals
5. ligands
6. covalent
7. hydrogen bonds
8. fatty acids
9. hydrolysis
10. cofactors
11. 1,000, or 10^3
12. oxidized
13. active site
14. polyunsaturated
15. kinases

B. Matching

1. F
2. E
3. T
4. A
5. R
6. X
7. S
8. N
9. B
10. G

C. True or False

The following items are false for the reasons stated. The other five are true.

1. Catabolism produces products with less free energy than the reactants.
6. The rate of clearance of a radioactive isotope from the body (biological half-life) depends on how fast its atoms decay (physical half-life) plus the rate at which the body excretes the isotope. Therefore, it cannot be longer than the physical half-life.
7. Polypeptides over 100 amino acids long are proteins; polysaccharides are chains of sugars, not amino acids.
8. Trans-fatty acids are more difficult to digest and therefore stay in the blood longer, leading to fat deposition in artery walls.
9. A particle of two or more atoms is a molecule even if the atoms are identical, as in O_2 and N_2.

D. Multiple Choice

1. d
2. a
3. b
4. e
5. c
6. a
7. d
8. e
9. a
10. c
11. a
12. a
13. c
14. d
15. d
16. d
17. a
18. b
19. b
20. b
21. e
22. a
23. b
24. d
25. d

E. Word Origins and Meanings

1. No, cut
2. Yes
3. No, heat
4. Yes
5. No, work
6. No, producing
7. Yes
8. Yes
9. No, part
10. Yes
11. many; polymer
12. joined; conjugated
13. sugar; glycogen
14. both; amphiphilic
15. to break down; catabolic

F. Which One Does Not Belong?

1. (d) steroids are made from cholesterol; the others are polymer → monomer catabolic reactions
2. (c) is a carbohydrate; others are lipids
3. (c) is not an effect of anabolic steroid use; the others are
4. (a) is not a reaction classification as the others are; many reactions are reversible
5. (c) is a monosaccharide; others are polysaccharides

Chapter 3: Cellular Form and Function

A. Short Answer

1. resolution
2. stellate
3. cytoplasm
4. glycolipids
5. transmembrane
6. aquaporins
7. brush border
8. nuclear pores
9. glycocalyx
10. secondary active transport
11. solutes
12. receptor-mediated endocytosis
13. cisterna
14. peroxisomes
15. Na^+–K^+ pump

B. Matching

1. M
2. N
3. D
4. X
5. R
6. S
7. T
8. O
9. F
10. E

C. True or False

The following items are false for the reasons stated. The other five are true.

2. In most cases, exocytosis is used to release a substance synthesized by the cell for use somewhere else, such as digestive enzymes.
3. Neither humans nor any other animals have cell walls. It's the plasma membrane and other cell membranes that are made of proteins and phospholipids.
6. Cilia and flagella have a 9+2 microtubular arrangement, but microvilli do not.
8. Louis Pasteur conducted historic experiments that *disproved* spontaneous generation for bacteria.
10. Cells cannot make energy but can only transfer it from one molecule to another or release it from chemical bonds. This is what the Na^+–K^+ pumps do to release heat.

D. Multiple Choice

1. d
2. c
3. a
4. a
5. b
6. e
7. a
8. b
9. b
10. a
11. e
12. b
13. a
14. d
15. e
16. c
17. d
18. a
19. b
20. c
21. e
22. b
23. e
24. a
25. a

E. Word Origins and Meanings

1. Yes
2. No, eat
3. No, spindle
4. Yes
5. Yes
6. Yes
7. No, reservoir
8. No, loosen, dissolve
9. No, push or thrust
10. Yes
11. power; dynein
12. formed; cytoplasm
13. first; protoplasm
14. characterized by; squamous
15. tension; hypertonic

F. Which One Does Not Belong?

1. (b) is not permanent and is a non-membranous organelle; others are permanent and have at least one membrane
2. (c) requires a membrane protein, others do not; however, (a) is the only type of transport that moves materials through an epithelium, not through the plasma membrane
3. (d) does not contain tubulin, as others do; (a) & (b) are made of microtubules
4. (d) Na^+ and K^+ are counter-transported in (a); (b) & (c) are functions of (a)
5. (a) is caused by a nuclear gene defect, others by mtDNA mutations

G. Figure Exercise

1. M	3. N	5. I	7. A	9. E
2. F	4. O	6. L	8. H	10. P

Chapter 4: Genetics and Cellular Function

A. Short Answer

1. exons
2. heterozygous
3. 2%, gene regulation
4. pleiotropy
5. polygenic inheritance
6. anticodon
7. tumor suppressor genes
8. recessive
9. growth factors
10. metastasize
11. rough ER cisternae
12. gene pool
13. DNA ligase
14. complementary base pairing
15. pyrimidines

B. Matching

1. B
2. A
3. L
4. V
5. X
6. S
7. C
8. E
9. W
10. I

C. True or False

The following items are false for the reasons stated. The other five are true.
4. Many proteins for use in the cytosol are synthesized by free ribosomes, not only by those on the rough ER.
5. mRNA does not have anticodons, but codons. Anticodons are found on tRNA.
7. A codon is found on mRNA, not on DNA, and it codes for only one amino acid, not a whole polypeptide. The statement would be true if the word *codon* is changed to *gene*.
8. Females inherit sex-linked traits if both their mother and father pass it down to them (thus, females can have color blindness, for example), but they are less common than in males.
10. Whether an allele is dominant or recessive has nothing to do with how common it is. For example, the most common blood type is type O, caused by a recessive allele.

D. Multiple Choice

1. a
2. b
3. c
4. d
5. a
6. a
7. c
8. c
9. a
10. c
11. a
12. c
13. c
14. a
15. d
16. b
17. d
18. d
19. c
20. e
21. c
22. c
23. c
24. b
25. a

E. Word Origins and Meanings

1. No, color
2. No, change
3. Yes
4. Yes
5. Yes
6. No, motion
7. No, new
8. No, bad
9. Yes
10. No, half
11. different; heterozygous
12. form; polymorphism
13. tumor; oncogene
14. apart; anaphase
15. fingers, toes; polydactyly

F. Which One Does Not Belong?

1. (a) is a stop codon on mRNA; others are codons for specific amino acids
2. (b) is a function of meiosis; others are accomplished by mitosis
3. (d) is having two different alleles for a given gene; others are types of inheritance wherein genes are expressed in different ways
4. (c) is a defense mechanism against cancer; others are environmental carcinogens
5. (d) is a tumor type encapsulated and slow-growing; by definition, *cancer* refers to a malignant tumor that spreads quickly

G. Genetics Application Problem

ANSWER: Color blindness is a sex-linked recessive trait. A female must have a double dose of recessive alleles to be colorblind. A male only needs one, since there is no allele for color blindness on the Y chromosome. Since Susan had normal color vision, she had to have one dominant allele (inherited from her mother); she had to be heterozygous for this trait, since we know her father had to give her his recessive allele. If Bill had normal vision, then he had a dominant allele and could not pass on anything but this dominant gene to his daughters. Bill could not be the father of a colorblind daughter. However, he could be the father of a colorblind boy with Susan as the mother. Half of her male children have a chance of being colorblind. See the text for the Punnett square and for sex-linked inheritance.

Susan: $X^B X^b$ Bill: $X^B Y$ Jennifer: $X^b X^b$

Chapter 5: Histology

A. Short Answer

1. chondrocytes
2. ectoderm
3. smooth
4. skeletal
5. blood or lymph
6. areolar (loose connective)
7. hyaline
8. macrophages
9. adipose, adipocytes
10. blood
11. holocrine
12. senile atrophy
13. goblet
14. tissue (interstitial) fluid
15. tissue engineering

B. Matching

1. X
2. S
3. U
4. M
5. Q
6. B
7. F
8. N
9. E
10. R

C. True or False

The following items are false for the reasons stated. The other five are true.
1. Transitional epithelium occurs in the urinary tract only.
3. Most embryonic stem cells come from donated IVF embryos.
5. There is no columnar epithelium in the heart—only simple squamous epithelium on its inner and outer surfaces.
7. Stratified squamous epithelium is the most durable type and, therefore, is found in places subject to stress and abrasion (skin, oral cavity, esophagus, anal canal, and vagina, for example).
8. The peritoneal cavity is lined by a serous membrane, the peritoneum, not by a mucous membrane.

D. Multiple Choice

1. c
2. c
3. a
4. b
5. e
6. d
7. a
8. a
9. b
10. c
11. d
12. c
13. e
14. a
15. e
16. c
17. a
18. b
19. e
20. e
21. e
22. d
23. a
24. c
25. c

E. Word Origins and Meanings

1. Yes
2. No, middle
3. No, network
4. No, form or produce
5. Yes
6. No, white
7. Yes
8. No, around
9. No, band
10. Yes
11. between; interstitial
12. nourishment; hypertrophy
13. away; apoptosis
14. glass; hyaline
15. horn; keratin

F. Which One Does Not Belong?

1. (c) is not a cell type in fibrous connective tissue, as are others
2. (b) is a dense connective tissue type; others are types of loose connective tissue
3. (d) is a primary tissue type; the others are connective tissues
4. (a) is a type of connective tissue; others are primary tissue types
5. (a) is dry (skin); others are moist membranes

G. Matching

1.   T
2.   N
3.   R
4.   H
5.   E
6.   I
7.   P
8.   F
9.   C
10. A
11. M
12. L
13. D
14. B
15. J

Chapter 6: The Integumentary System

A. Short Answer

1. corneum
2. keratinocytes
3. Merkel/tactile
4. desquamate
5. dermal papillae
6. hematoma
7. vitamin D (calcitriol)
8. vasoconstriction
9. terminal
10. medulla
11. debridement
12. nail matrix
13. sebaceous
14. second-degree
15. melanoma

B. Matching

1. E
2. H
3. R
4. C
5. N
6. S
7. L
8. O
9. P
10. F

C. True or False

The following items are false for the reasons stated. The other five are true.
1. No skin cancer is benign; cancer by definition is malignant.
4. Jaundice is caused by bilirubin, usually as a result of liver disease or hemolytic anemia.
5. The epidermis is made of epithelium. The dermis is made of fibroconnective tissue.
8. The body hair of children is mostly vellus; lanugo is usually limited to the fetus.
9. Hair growth results from mitosis in the hair bulb.

D. Multiple Choice

1. d
2. c
3. c
4. e
5. b
6. d
7. a
8. c
9. b
10. a
11. c
12. c
13. a
14. a
15. e
16. a
17. b
18. c
19. d
20. e
21. c
22. c
23. d
24. a
25. b

E. Word Origins and Meanings

1. Yes
2. No, horn
3. Yes
4. Yes
5. Yes
6. No, vessel
7. No, sweat
8. Yes
9. Yes
10. No, nail
11. tree, branch; dendritic
12. black; melanin
13. mass; hematoma
14. down; catagen
15. shaggy; hirsutism

F. Which One Does Not Belong?

(a) not found in stratum basale in epidermis, as are others
(b) not an area for growth/mitosis, as others are
(d) produces earwax; others are sweat glands
(c) is the outer layer of the epidermis in skin; others are associated with nails
(a) is a function of hair; others are skin functions

G. Figure Exercise

1. Q
2. N (or T)
3. V
4. X
5. E
6. J
7. O
8. G
9. U
10. W

Chapter 7: Bone Tissue

A. Short Answer

1. osteocalcin
2. axial
3. canaliculi
4. compact, spongy
5. periosteum
6. osteoblasts
7. central canal
8. red bone marrow (myeloid tissue)
9. acid phosphatase
10. rickets/osteomalacia, osteogenesis imperfecta
11. epiphyseal plate
12. Wolff's law
13. hypocalcemia, tetany
14. parathyroid hormone, osteoclasts
15. osteoporosis

B. Matching

1. H	3. C	5. M	7. O	9. A
2. I	4. A	6. F	8. R	10. X

C. True or False

The following items are false for the reasons stated. The other five are true.

2. Bones do not attain their maximum mass and density until middle age. They undergo mineral loss in old age and are continually remodeled in response to gravity and muscular tension throughout life.
4. The epiphyseal plate closes by the end of adolescence; it makes no contribution to bone growth in adults.
5. The trabeculae are arranged along lines of stress, not at random.
6. Calcitonin has little effect in most adults.
9. Most people produce adequate amounts of vitamin D internally, as long as they have enough daily exposure to sunlight.

D. Multiple Choice

1. e	6. a	11. b	16. e	21. e
2. d	7. a	12. a	17. d	22. c
3. c	8. e	13. b	18. c	23. a
4. e	9. d	14. a	19. b	24. b
5. d	10. d	15. e	20. a	25. e

E. Word Origins and Meanings

1. Yes
2. No, resembling
3. No, growth
4. Yes
5. No, forming
6. Yes
7. Yes
8. Yes
9. Yes
10. No, straight
11. form, produce; osteoblast
12. upon, above; epiphysis
13. little; trabeculae
14. around; periosteum
15. cartilage; chondrocyte

F. Which One Does Not Belong?

1. (b) is the organic matrix of bone secreted by osteogenic cells; others are cell types found in bone
2. (c) is inorganic crystal (calcium phosphate) in bone; others are calcium-regulating hormones
3. (b) all others are long bones except the tarsals that are short bones
4. (d) "straight" is not a fracture type, as are the others
5. (a) is a step in endochondral bone formation; others are steps in bone fracture repair

G. Figure Exercise

1. d
2. a
3. a
4. d
5. b

6. c
7. c
8. a
9. b
10. d

Chapter 8: The Skeletal System

A. Short Answer

1. sutures
2. foramina
3. palatine, maxillae
4. sphenoid
5. infraspinous fossa
6. ulna
7. intervertebral discs
8. intervertebral foramina
9. xiphoid process
10. acromion
11. sacrum, and hip bones
12. pubic symphysis
13. ischial tuberosities
14. malleoli
15. phalanges

B. Matching

1. D
2. E
3. U
4. M
5. L
6. P
7. S
8. I
9. C
10. J

C. True or False

The following items are false for the reasons stated. The other five are true.

1. The coronal suture does this. The lambdoid suture separates the parietal bones from the occipital bone.
2. It originates on the temporal lines of the parietal bone.
3. The fibula is lateral.
6. Bifid spinous processes are characteristic of the cervical vertebrae, not of the thoracics.
7. The zygomatic arch is made by the articulation of the zygomatic process of the temporal bone and the temporal process of the zygomatic bone.

D. Multiple Choice

1. c
2. d
3. c
4. e
5. e
6. b
7. b
8. d
9. c
10. a
11. b
12. b
13. d
14. c
15. d
16. c
17. c
18. d
19. d
20. b
21. c
22. b
23. e
24. a
25. c

E. Word Origins and Meanings

1. No, tough
2. Yes
3. No, time
4. No, breast
5. Yes
6. No, spade, shovel
7. Yes
8. No, crest
9. Yes
10. No, rib
11. layer, plate; laminectomy
12. hammer, club, key; clavicle
13. above; supraorbital foramen
14. before; antebrachium
15. head; capitulum

F. Which One Does Not Belong?

1. (d) is found on the humerus; the rest are on the ulna
2. (c) is on the occipital bone; others are found on the temporal bone
3. (d) is an articulation site with the humerus; others are muscle attachment sites
4. (d) is on the humerus; the others are found on the femur
5. (a) is the only axial skeleton bone and is the only flat bone listed

G. Matching Articulations and Bone Markings

I.	1.	A	6.	J	11.	P
	2.	M	7.	G	12.	I
	3.	K	8.	O	13.	C
	4.	D	9.	B	14.	E
	5.	R	10.	F	15.	H

II.	1.	C	7.	F	13.	B	19.	C
	2.	E	8.	E	14.	A	20.	F
	3.	C	9.	C	15.	C	21.	A
	4.	E	10.	B	16.	E	22.	C
	5.	A	11.	A	17.	B	23.	A
	6.	E	12.	B	18.	B	24.	F

Chapter 9: Joints

A. Short Answer

1. synovial
2. frontal
3. interosseous membrane
4. tendon, ligament
5. condylar
6. pivot
7. flexing
8. plantar flexion
9. supination
10. range of motion
11. first-class
12. shoulder (glenohumeral)
13. musculoskeletal movement
14. knee
15. bursae

B. Matching

1. G
2. X
3. W
4. A
5. R
6. H
7. O
8. I
9. V
10. K

C. True or False

The following items are false for the reasons stated. The other five are true.
3. A cartilaginous joint can be either an amphiarthrosis or a synarthrosis, but not a diarthrosis.
5. The mandible must be protracted to bring the incisors in line in order to take a bite.
8. One of them, the subscapularis, is on the anterior side.
9. There is no periosteum over an articular cartilage.
10. Hinge joints are monaxial. Biaxial joints can move in two planes, whereas flexion and extension of a hinge joint are just opposite movements in a single plane.

D. Multiple Choice

1. c
2. e
3. a
4. e
5. e
6. b
7. a
8. a
9. d
10. c
11. a
12. d
13. d
14. b
15. c
16. b
17. a
18. d
19. c
20. e
21. d
22. e
23. d
24. a
25. e

E. Word Origins and Meanings

1. No, joint
2. Yes
3. No, together
4. Yes
5. Yes
6. No, to lay back
7. Yes
8. No, round
9. No, cross
10. Yes
11. purse; bursa
12. on all sides; amphiarthrosis
13. nail (bolt); gomphosis
14. growth; symphysis
15. little; meniscus

F. Which One Does Not Belong?

1. (d) is a bony joint formed by the fusion of two bones; others are parts of a synovial joint
2. (b) is a fibrocartilaginous joint; others are synovial joints
3. (c) is a synovial joint; the rest are fibrous joints
4. (d) is a ligament in the hip joint; others are all knee ligaments
5. (a) is not a movement of the foot, as the others are

G. Figure Exercise

1. E	6. D	11. B
2. G	7. C	12. J
3. F	8. F	13. D
4. A	9. H	14. A
5. B	10. G	15. E

Chapter 10: The Muscular System

A. Short Answer

1. sphincter
2. modiolus
3. fusiform
4. prime mover, antagonist
5. intrinsic, extrinsic
6. zygomaticus
7. temporalis, masseter
8. fixators
9. external intercostals, diaphragm
10. rectus abdominis
11. quadriceps femoris
12. deltoid
13. rotator cuff
14. brachialis
15. cranial, spinal

B. Matching

1. D	3. S	5. B	7. J	9. F
2. R	4. Q	6. G	8. N	10. V

C. True or False

The following items are false for the reasons stated. The other five are true.
3. It is formed by the mylohyoid.
4. It is the trapezius.
5. The transverse abdominal is deep to both of the abdominal obliques.
7. It is on the anterior side of the scapula. The muscle on the posterior side, in the position described, is the infraspinatus.
8. These two muscles are antagonists.

D. Multiple Choice

1. b	6. b	11. a	16. d	21. a
2. c	7. d	12. c	17. e	22. c
3. b	8. d	13. a	18. c	23. b
4. d	9. e	14. b	19. a	24. c
5. e	10. a	15. b	20. b	25. d

E. Word Origins and Meanings

1. Yes	6. Yes	11. clavicle; sternocleidomastoid
2. Yes	7. Yes	12. straight; rectus abdominis
3. Yes	8. No, neck	13. tailor; sartorius
4. No, lip	9. Yes	14. cheek; buccinator
5. No, mouth	10. Yes	15. partition; diaphragm

F. Which One Does Not Belong?

1. (c) is an extrinsic muscle of the hand; the others are intrinsic
2. (d) is not a hamstring, as are others
3. (a) pertains to number of heads; others are location names
4. (b) is not a general function of muscles but a specific action; others are general functions
5. (c) is a suprahyoid group muscle and opens the mouth; others are muscles of mastication and close the mouth

G. Figure Exercises

I.
1. semitendinosus, semimembranosus, biceps femoris
2. trapezius
3. supraspinatus, infraspinatus, teres minor; subscapularis is not shown in this view
4. triceps brachii
5. serratus anterior
6. gastrocnemius
7. semitendinosus and semimembranosus
8. lateral and medial heads, triceps brachii; flexor carpi ulnaris; extensor digitorum
9. gluteus maximus
10. latissimus dorsi

II.

1.	A	6.	K	11.	D
2.	E	7.	O	12.	Q
3.	B	8.	L (P)	13.	R
4.	J	9.	N	14.	F
5.	M	10.	G	15.	H

III.

1. subscapularis
2. infraspinatus
3. supraspinatus
4. teres minor
5. teres major
6. deltoid
7. trapezius
8. levator scapulae
9. rhomboideus major and minor
10. serratus anterior
11. pectoralis minor
12. biceps brachii (both heads)
13. triceps brachii (long head only)
14. latissimus dorsi
15. coracobrachialis

Chapter 11: Muscle Cells

A. Short Answer

1. T tubules
2. terminal cisternae of SR
3. myosin
4. autorhythmic
5. neuromuscular junction
6. acetylcholine
7. resting membrane potential
8. action potential
9. sliding filament
10. troponin, tropomyosin
11. wave summation (incomplete tetanus)
12. eccentric
13. oxygen debt
14. single-unit
15. norepinephrine

B. Matching

1. I	3. A	5. M	7. N	9. G
2. R	4. U	6. E	8. S	10. L

C. True or False

The following items are false for the reasons stated. The other five are true.

1. Myofilaments slide over each other, shortening the sarcomere but do not individually shorten during muscle contraction.
4. One muscle fiber is never supplied by more than one nerve fiber. A large motor unit is one in which each nerve fiber supplies many muscle fibers (opposite from the way the statement is worded).
5. Myofibrils are organelles made up of myofilaments arranged into sarcomeres.
6. Myosin can bind only ATP. The role of creatine phosphate is to donate phosphate groups to make more ATP in times of oxygen deficit.
10. Most of the A band is a region of overlapping myosin and actin.

D. Multiple Choice

1. d	6. a	11. b	16. a	21. c
2. b	7. a	12. a	17. b	22. a
3. c	8. c	13. e	18. c	23. d
4. c	9. c	14. c	19. e	24. b
5. d	10. b	15. d	20. e	25. c

E. Word Origins and Meanings

1. No, flesh or muscle	6. No, abnormal	11. husk; sarcolemma
2. Yes	7. Yes	12. of death; rigor mortis
3. Yes	8. Yes	13. growth; dystrophy (muscular dystrophy)
4. No, length	9. No, staircase	14. self; autorhythmic
5. No, tension	10. Yes	15. precursor; myoblast

F. Which One Does Not Belong?

1. (c) is not part of a neuromuscular junction, as the others are; the SR is part of a muscle cell
2. (d) is not a characteristic of all muscle types; only smooth and cardiac muscle are autorhythmic
3. (a) all are calcium-binding proteins except tropomyosin
4. (a) is used in cross-bridge cycling in smooth muscle; others are means of making ATP for muscle contraction
5. (b) is not found in cardiac muscle, as others are

G. Figure Exercises

1.   c
2.   d
3.   b
4.   c (#24 is also a z-disc)
5.   e
6.   c
7.   a
8.   d
9.   c
10. e
11. e
12. b
13. d
14. a
15. c

Chapter 12: Nerve Cells

A. Short Answer

1. glial
2. dendrites, axon
3. trigger zone
4. efferent (motor)
5. absolute refractory period
6. oligodendrocytes, Schwann cells
7. nodes of Ranvier
8. synaptic vesicles
9. depolarization
10. excitatory postsynaptic potential
11. threshold
12. neural integration
13. satellite cells
14. epinephrine, norepinephrine, dopamine
15. presynaptic inhibition

B. Matching

1. B	3. X	5. M	7. J	9. N
2. Q	4. L	6. T	8. U	10. S

C. True or False

The following items are false for the reasons stated. The other five are true.

1. Each node slows down the signal a little; myelinated nerve fibers conduct signals faster because of the internodes, not the nodes.
2. In the PNS, neurosomas are located in the ganglia.
4. Parkinsonism is caused by degeneration of dopamine neurons in the substantia nigra of the brain.
5. All action potentials have the same voltage regardless of stimulus strength, according to the all-or-none law.
9. cAMP activates protein kinases. It does not directly open membrane ion gates.

D. Multiple Choice

1. a	6. a	11. d	16. c	21. c
2. b	7. a	12. a	17. c	22. b
3. a	8. e	13. d	18. b	23. a
4. c	9. b	14. c	19. c	24. b
5. e	10. e	15. c	20. a	25. e

E. Word Origins and Meanings

1. Yes	6. Yes	11. body; somatic motor
2. No, toward	7. Yes	12. tree; arborization
3. No, carry	8. No, touch	13. few; oligodendrocyte
4. Yes	9. Yes	14. to leap; saltatory
5. No, star	10. No, change	15. out, away from; efferent

F. Which One Does Not Belong?

1. (a) is in an axon terminal; the rest are in the neurosoma of a neuron
2. (c) neuron type that has no axons; all others have them
3. (a) part of the CNS; others are all parts of the PNS
4. (d) neuroglia in the PNS; others are glial cells in the CNS
5. (a) is characteristic of an action potential; all others are characteristic of a graded potential

G. Figure Exercise

1. e
2. e
3. c
4. a
5. e
6. b
7. d
8. a
9. e
10. a

Chapter 13: The Spinal Cord, Spinal Nerves, and Somatic Reflexes

A. Short Answer

1. ganglia
2. epineurium
3. spinal
4. posterior horn
5. gracilis, cuneatus
6. decussation
7. tracts
8. brachial
9. gamma motor, alpha motor
10. monosynaptic
11. dura mater, arachnoid mater, pia mater
12. posterior
13. white matter
14. subarachnoid space
15. medullary cone (conus medullaris)

B. Matching

1. C	3. K	5. V	7. L	9. U
2. E	4. N	6. S	8. X	10. J

C. True or False

The following items are false for the reasons stated. The other five are true.

1. The lateral horns contain neurosomas of ANS neurons whose fibers pass out the anterior root at the same cord level.
2. Most nerves carry both sensory and motor fibers and are, therefore, mixed nerves.
5. Because most fibers decussate, the origins and destinations are on opposite or contralateral sides of the body.
7. Gamma motor neurons adjust tension in the muscle spindle; alpha motor neurons stimulate muscle contraction.
10. It is an ascending tract that carries information about light touch, pain, pressure, and temperature.

D. Multiple Choice

1. c	6. c	11. d	16. e	21. c
2. c	7. c	12. a	17. e	22. c
3. d	8. d	13. a	18. a	23. d
4. e	9. c	14. a	19. b	24. e
5. b	10. e	15. e	20. a	25. e

E. Word Origins and Meanings

1. No, knot
2. Yes
3. Yes
4. Yes
5. No, section or segment
6. Yes
7. No, stretch
8. Yes
9. No, pressure
10. No, feeling
11. resembling; subarachnoid
12. tough; dura mater
13. gray matter; poliomyelitis
14. one's own; proprioception
15. forked, divided; spina bifida

F. Which One Does Not Belong?

1. (b) is in the spinal cord (CNS); the rest are part of the PNS
2. (d) has multiple causes; all others are caused by viruses
3. (c) the ansa cervicalis from the cervical plexus carries motor fibers only; others are mixed nerves derived from the brachial plexus
4. (a) is a descending tract; all others are ascending pathways
5. (d) is caused by astrocyte malfunction leading to toxic levels of glutamate in synapses; others are caused by various viral infections

G. Figure Exercise

1.	c	6. a
2.	d	7. b
3.	c	8. a
4.	e	9. c
5.	c	10. d

Chapter 14: The Brain and Cranial Nerves

A. Short Answer

1. gray matter
2. meninges
3. dura mater
4. blood–brain barrier
5. facial (CN VII)
6. cingulate gyrus
7. vagus (CN X)
8. cerebellum
9. substantia nigra
10. corpus callosum
11. occipital
12. orexins
13. limbic system
14. beta
15. association area

B. Matching

1. P
2. C
3. D
4. X
5. V
6. S
7. N
8. Q
9. F
10. H

C. True or False

The following items are false for the reasons stated. The other five are true.
2. The optic nerve functions in vision; CN III, IV, and VI control movements of the eyeball and eyelid.
5. The neural folds and neural groove are early embryonic structures. The folds and grooves of the mature cerebrum are sulci and giri.
6. The choroid plexuses secrete the CSF, and the arachnoid villi reabsorb it.
9. Purkinje cells are found in the cerebellum. The distinctive neurons of the cerebral cortex are pyramidal cells.
10. EEGs are composite recordings of postsynaptic potentials in the cerebral cortex, not of action potentials.

D. Multiple Choice

1. a
2. e
3. a
4. a
5. c
6. e
7. c
8. b
9. d
10. a
11. e
12. a
13. c
14. d
15. e
16. b
17. c
18. a
19. a
20. a
21. e
22. c
23. b
24. d
25. a

E. Word Origins and Meanings

1. No, little
2. Yes
3. Yes
4. No, brain
5. Yes
6. Yes
7. No, cross over
8. Yes
9. No, rind or bark
10. No, girdle
11. end, remote; telencephalon
12. spinal cord; myelencephalon
13. worm; vermis
14. lens; lentiform nucleus
15. knowledge; agnosia

F. Which One Does Not Belong?

1. (b) is in the hypothalamus; the others are in the medulla oblongata
2. (d) is part of the brain stem; the rest are parts of the midbrain
3. (a) is related to sensory perception of touch, pain, pressure, etc.; others are brain areas associated with speech
4. (c) uses primarily motor pathways; the others use primarily sensory pathways
5. (b) requires greater use of the categorical hemisphere, while the others require the representational hemisphere

G. Matching

I. 1. A   4. E   7. B   10. C   13. H
    2. D   5. E   8. E   11. H   14. G
    3. E   6. F   9. I, C   12. H

II. 1. I   6. G   11. K   16. J   21. E
    2. L   7. K   12. M   17. L   22. A
    3. I   8. M   13. N   18. L   23. D
    4. H   9. N   14. C   19. F   24. K
    5. I   10. J   15. C   20. K   25. B

Chapter 15: The Autonomic Nervous System and Visceral Reflexes

A. Short Answer

1. sympathetic, parasympathetic
2. communicating rami
3. preganglionic, postganglionic
4. nicotinic, muscarinic
5. postganglionic sympathetic neurons
6. cranial
7. cholinergic
8. dual innervation
9. sympathetic tone
10. hypothalamus
11. excitatory, inhibitory
12. chain
13. parasympathetic division
14. preganglionic
15. splanchnic

B. Matching

1. G	3. K	5. H	7. S	9. D
2. I	4. R	6. F	8. P	10. N

C. True or False

The following items are false for the reasons stated. The other five are true.

4. This occurs when norepinephrine binds beta-adrenergic receptors on vascular smooth muscle in these tissues.
5. Most organs are innervated by BOTH divisions and, therefore, have dual innervation.
6. Caffeine blocks the NT adenosine that makes us sleepy.
8. The heart has its own pacemaker and can beat on its own; however, the ANS adjusts heart rate up or down.
10. Somatic motor fibers are myelinated; postganglionic ANS fibers are unmyelinated.

D. Multiple Choice

1. d	6. a	11. b	16. d	21. c
2. b	7. c	12. a	17. c	22. a
3. d	8. b	13. d	18. d	23. e
4. a	9. b	14. c	19. a	24. e
5. b	10. a	15. e	20. d	25. b

E. Word Origins and Meanings

1. Yes
2. Yes
3. No, rule
4. No, feeling
5. Yes
6. Yes
7. Yes
8. No, near
9. No, intestine
10. No, breakdown, destroy
11. self; autonomic
12. branch; communicating ramus
13. together, sympathetic
14. wall; intramural ganglia
15. imitate; sympathomimetics

F. Which One Does Not Belong?

1. (c) is the NT in the sympathetic division; others are associated with the parasympathetic division
2. (a) receives somatic motor innervation; (b) and (d) are involuntary and are under ANS control
3. (c) is a parasympathetic ganglion; others are in the sympathetic division
4. (b) is associated with the sympathetic division; others are associated with the parasympathetic division
5. (a) is under control of the parasympathetic division; others are under sympathetic control

G. Identification Exercise

1. S	6. P	11. S	16. P	21. S
2. P	7. P	12. S	17. P	22. P
3. P	8. S	13. S	18. S	23. P
4. P	9. P	14. P	19. S	24. S
5. S	10. P	15. S	20. S	25. S

Chapter 16: Sense Organs

A. Short Answer

1. proprioceptors
2. fast pain
3. vallate
4. olfactory bulbs
5. frequency
6. malleus
7. cochlea
8. semicircular ducts
9. lacrimal
10. foliate
11. macula utriculi
12. outer hair cells
13. aqueous humor
14. fovea centralis
15. retinal

B. Matching

1. L
2. R
3. V
4. K
5. F
6. B
7. O
8. X
9. G
10. C

C. True or False

The following items are false for the reasons stated. The other five are true.

2. The posterior chamber is between the iris and lens, not the space behind the lens; it is filled with aqueous humor.
3. Each cone has only one type of pigment, but there are three types of cones with different absorption spectra.
4. The auditory ossicles do not change the vibration frequency.
6. Substance P is the neurotransmitter that transmits pain signals; enkephalin blocks them.
10. Hair cells of the semicircular ducts work this way, but not those in the cochlea. Their stereocilia are bent as the basilar membrane vibrates up and down and the tectorial membrane remains relatively still.

D. Multiple Choice

1. d
2. c
3. c
4. a
5. e
6. c
7. e
8. a
9. c
10. a
11. b
12. c
13. b
14. e
15. b
16. c
17. e
18. b
19. b
20. c
21. b
22. c
23. c
24. b
25. a

E. Word Origins and Meanings

1. Yes
2. Yes
3. No, drum
4. Yes
5. Yes
6. No, ear
7. Yes
8. No, stone
9. No, dark
10. No, pit
11. leaf; foliate
12. process, condition; otosclerosis
13. spot; macula sacculi
14. solid; stereocilia
15. bearer; chromatophore

F. Which One Does Not Belong?

1. (d) is not part of the cochlea as the others are; also, (d) is a perilymph-filled chamber, whereas others are not chambers
2. (a) is a receptor modality; others are receptor classes based on the origin of stimuli
3. (d) is in the vestibular organ; others are part of hearing apparatus
4. (b) is an unencapsulated nerve ending; others are encapsulated
5. (b) is the normal state of a relaxed eye focused on distant objects; others are visual defects

G. Figure Exercise

1.	N	6.	O	11.	O	16.	N
2.	D	7.	H	12.	B	17.	C
3.	I	8.	A	13.	Q	18.	D
4.	B	9.	C	14.	L	19.	M
5.	F	10.	G	15.	I	20.	D

Chapter 17: The Endocrine System

A. Short Answer

1. neuroendocrine cells
2. TRH
3. hypophyseal portal system
4. negative feedback inhibition
5. gigantism, acromegaly
6. thyroid
7. glucocorticoids, mineralocorticoids
8. pancreatic islets
9. glucocorticoids
10. testosterone
11. thyroxine-binding globulin
12. extracellular, intracellular
13. second messengers
14. glucosuria, hyperglycemia
15. enzyme amplification

B. Matching

1. G	3. D	5. R	7. H	9. I
2. E	4. A	6. T	8. P	10. M

C. True or False

The following items are false for the reasons stated. The other five are true.

2. The receptors for steroid hormones are in the nucleus.
3. Polypeptides do not enter their target cells. They remain bound to the cell surface, and second messengers within the cell trigger various metabolic responses.
5. These hormones are synthesized in hypothalamic nuclei, transported by axoplasmic flow, and stored and released by the posterior pituitary.
6. Their response is slow (hours to days) because these hormones activate genes; transcription and translation take time.
8. Insulin is secreted when blood glucose level rises; its effect is to lower blood glucose concentration.

D. Multiple Choice

1. b	6. d	11. d	16. a	21. a
2. a	7. c	12. d	17. a	22. d
3. a	8. e	13. a	18. b	23. c
4. e	9. b	14. e	19. a	24. a
5. e	10. c	15. b	20. e	25. a

E. Word Origins and Meanings

1. No, gland	6. Yes	11. secrete; endocrine
2. Yes	7. No, flow through	12. milk; prolactin
3. No, favoring	8. Yes	13. to pass through; antidiuretic
4. No, turn or change	9. Yes	14. drinking; polydipsia
5. No, birth	10. Yes	15. honey; diabetes mellitus

F. Which One Does Not Belong?

1. (c) GH *is* somatotropin; others are correct anterior pituitary hormone–target gland relationships
2. (d) is an amine hormone; others are protein in nature
3. (a) is the only hypoglycemic agent; the others increase blood glucose concentration
4. (d) breaks down the second messenger cAMP; others are second messengers
5. (c) is not an anterior pituitary hormone, as others are; PTH comes from parathyroid glands and increases calcium concentration

G. Matching

1. F	3. J	5. C	7. G	9. D
2. A	4. B	6. I	8. H	10. E

Chapter 18: The Circulatory System: Blood

A. Short Answer

1. viscosity
2. hemopoiesis, hemopoietic
3. leukocytes
4. macrophages
5. heme group
6. hematocrit/packed cell volume
7. primary polycythemia
8. A, O
9. Rh−, Rh+
10. neutrophils
11. acute myelocytic leukemia
12. hemostasis
13. extrinsic, tissue thromboplastin
14. fibrinogen, fibrin
15. thrombosis, embolus

B. Matching

1. A
2. R
3. O
4. S
5. F
6. V
7. J
8. C
9. M
10. K

C. True or False

The following items are false for the reasons stated. The other five are true.
3. An oxygen deficiency is a result of anemia, not the cause.
4. Bile pigments come from breakdown of the heme moiety.
6. A person with type B blood has anti-A antigens; a person with type A blood has anti-B antigens.
7. Although sodium ions contribute to blood osmolarity, they are not colloidal; colloid osmotic pressure is produced by the proteins of the blood plasma.
8. Most of the blood plasma is water.

D. Multiple Choice

1. d
2. a
3. e
4. c
5. b
6. c
7. c
8. e
9. e
10. c
11. a
12. b
13. b
14. a
15. b
16. d
17. e
18. d
19. a
20. a
21. e
22. e
23. d
24. b
25. c

E. Word Origins and Meanings

1. Yes
2. No, red
3. No, formation
4. Yes
5. Yes
6. Yes
7. Yes
8. No, increase
9. No, stay or remain
10. No, mass
11. white; leukemia
12. green, biliverdin
13. red; bilirubin
14. clotting; thrombopoietin
15. to separate; hematocrit

F. Which One Does Not Belong?

1. (d) is a cell fragment; others are leukocytes
2. (c) is carried by hemoglobin; others are essential in hemoglobin synthesis
3. (b) is an unwanted clot in an unbroken vessel; others are hemostatic mechanisms
4. (d) is a nitrogen-containing nutrient in plasma; others are nitrogenous wastes excreted by kidneys
5. (a) is not a correct relationship between precursor cells and formed elements; megakaryocytes are fragmented to become platelets

Chapter 19: The Circulatory System: The Heart

A. Short Answer

1. systemic, pulmonary
2. pericardium
3. myocardial infarction
4. papillary, tendinous cords
5. great cardiac, middle cardiac
6. ventricular fibrillation
7. sarcoplasmic reticulum, extracellular fluid
8. intercalated discs
9. sinoatrial node
10. pulmonary
11. calcium
12. T
13. systole, diastole
14. stroke volume, ejection fraction
15. chronotropic, inotropic

B. Matching

1. A
2. K
3. H
4. P
5. E
6. V
7. U
8. X
9. M
10. W

C. True or False

The following items are false for the reasons stated. The other five are true.
1. Afterload opposes the emptying of a heart chamber. Therefore, increasing afterload reduces the ejection fraction.
3. Cells throughout the myocardium will exhibit rhythmic spontaneous depolarizations if allowed to, but their rhythm is too slow to sustain life for long.
6. Athletes' hearts are larger and, thus, have higher stroke volumes, so, in order to produce the same cardiac output, they do not have to beat as fast. Their resting heart rates are, therefore, lower than average.
8. Both ventricles have equal output; this prevents pulmonary or systemic edema.
9. The atrioventricular valves, not the semilunar valves, do this.

D. Multiple Choice

1. b
2. d
3. a
4. d
5. c
6. d
7. c
8. d
9. e
10. a
11. d
12. d
13. a
14. b
15. d
16. b
17. c
18. d
19. a
20. d
21. e
22. d
23. b
24. c
25. b

E. Word Origins and Meanings

1. Yes
2. No, around
3. No, lower part
4. Yes
5. No, pertaining to
6. No, stuffed
7. Yes
8. Yes
9. Yes
10. No, fast
11. narrow; valvular stenosis
12. little beam; trabeculae carneae
13. pulse; sphygmomanometer
14. to choke, strangle; angina pectoris
15. hardening; atherosclerosis

F. Which One Does Not Belong?

1. (c) is an uncontrollable risk factor for coronary artery disease; others are controllable risk factors
2. (d) is the structure into which coronary veins drain; others carry arterial blood to cardiocytes
3. (a) anchor the AV valve flaps; the others are parts of the cardiac conduction system
4. (b) is regulated by cardiac centers in the brain and by the ANS, ion concentration, etc.; the others affect stroke volume (which, with heart rate, determines cardiac output)
5. (d) ventricular repolarization is measured by the T wave; other relationships in the EKG are correct

G. Figure Exercise

I. 1. A, D, O
    2. P, R
    3. I, U
    4. E
    5. Q
    6. F
    7. V
    8. A
    9. C
    10. E, K

II. 1. F   9. J
     2. H   10. D
     3. O   11. C
     4. E   12. I
     5. M   13. A
     6. L   14. G
     7. K   15. P
     8. N   16. B

Chapter 20: The Circulatory System: Blood Vessels and Circulation

A. Short Answer

1.	portal system	9.	oncotic
2.	tunica interna, endothelium	10.	venous pooling shock
3.	capillaries	11.	common iliac arteries
4.	medullary ischemic	12.	left subclavian
5.	systolic, diastolic	13.	superior mesenteric
6.	conducting, distributing	14.	brachiocephalic
7.	resistance, radius	15.	hepatic portal
8.	autoregulation		

B. Matching

1. X	3. F	5. U	7. A	9. R
2. T	4. I	6. M	8. K	10. C

C. True or False

The following items are false for the reasons stated. The other five are true.
2. An aneurysm is a weak, bulging vessel with a high likelihood of rupturing.
4. It is a vasoconstrictor.
6. Blood capillaries reabsorb approximately 85% of the fluid they emit and the lymphatic system absorbs the rest.
8. Arteries have no valves.
9. Blood flow increases 16-fold when the radius doubles.

D. Multiple Choice

1. d	6. b	11. a	16. c	21. a
2. d	7. c	12. e	17. b	22. c
3. c	8. b	13. d	18. c	23. c
4. a	9. d	14. a	19. a	24. e
5. e	10. e	15. a	20. a	25. b

E. Word Origins and Meanings

1.	No, added on	6.	No, pressure	11.	layer; laminar flow
2.	No, vessel	7.	Yes	12.	tip; thoracoacromial
3.	Yes	8.	No, neck	13.	rib; intercostal
4.	No, next	9.	No, stupor	14.	union, mate; azygos
5.	Yes	10.	No, diaphragm	15.	belly; celiac trunk

F. Which One Does Not Belong?

1. (c) is a paired artery; others are unpaired arteries
2. (d) is not a means of capillary exchange, as the others are
3. (a) is the way elastic arteries smooth out fluctuations in pressure between systole and diastole; others are factors affecting peripheral resistance
4. (d) is not a mechanism directly affecting venous return, as the others are
5. (b) decreases blood pressure; the others increase blood pressure by various mechanisms

G. Blood Vessel Exercises

I.
1. right axillary a.
2. right femoral v.
3. superior vena cava
4. basilar a.
5. left external and internal iliac aa.
6. right femoral v.
7. right common carotid a. and right subclavian a.
8. superior vena cava
9. left gastric a., splenic a., common hepatic a.
10. hepatic sinusoids to hepatic veins

II.
1. brachiocephalic trunk
2. left common carotid a.
3. left subclavian a.
4. posterior intercostal aa.
5. bronchial a.
6. esophageal aa.
7. celiac trunk
8. superior mesenteric a.
9. renal aa.
10. gonadal aa.
11. lumbar aa.
12. inferior mesenteric a.
13. common iliac aa.
14. inferior phrenic a.
15. middle suprarenal a.
16. median sacral a.

Chapter 21: The Lymphatic and Immune Systems

A. Short Answer

1. subclavian
2. pathogens
3. the tonsils
4. respiratory burst
5. inflammation
6. heparin, histamine
7. interferons
8. lymph nodes
9. cortisol
10. antigen-presenting
11. opsonization
12. nonspecific resistance
13. agglutination
14. cytotoxic T (T_c)
15. anaphylactic shock

B. Matching

1. O
2. D
3. F
4. I
5. A
6. Q
7. S
8. H
9. L
10. T

C. True or False

The following items are false for the reasons stated. The other five are true.
1. Highly repetitive molecules are usually not antigenic; antigens are structurally complex molecules that can be unique to each individual.
6. Opsonization makes bacteria easier to phagocytize. The process described in the statement is cytolysis.
8. Plasma cells have nothing to do with producing blood plasma; they produce circulating antibodies.
9. Blood pressure is dangerously *low* in anaphylactic shock.
10. Both the heavy and light chains have a genetically constant part (C region) and a variable part (V region).

D. Multiple Choice

1. e
2. c
3. a
4. b
5. e
6. b
7. d
8. c
9. b
10. c
11. e
12. a
13. b
14. e
15. d
16. d
17. d
18. e
19. c
20. c
21. c
22. e
23. d
24. e
25. c

E. Word Origins and Meanings

1. No, disease
2. No, freedom
3. Yes
4. Yes
5. Yes
6. No, between
7. No, altered
8. No, back
9. Yes
10. No, prevention
11. producing; pathogen
12. split apart, break down; cytolysis
13. disease; pathogen
14. to prepare food; opsonization
15. inflammation; lymphadenitis

F. Which One Does Not Belong?

1. (c) is not a function of the spleen, as the others are
2. (a) destroy cancer cells or virus-infected cells; others are macrophages with various functions
3. (b) is a function of cellular immunity; others are functions of inflammation
4. (c) synthesizes antibodies in humoral immunity; the others are T cells involved in cellular immunity
5. (b) involves movement of leukocytes out of capillaries in inflammation; others are mechanisms of humoral immunity

Chapter 22: The Respiratory System

A. Short Answer

1. respiratory, conducting
2. septum, fossae
3. epiglottis, vestibular folds
4. secondary bronchus
5. bronchioles
6. alveoli
7. ventral respiratory group
8. pleurae
9. surfactant
10. spirometer
11. dyspnea, hyperpnea
12. Boyle's
13. Henry's
14. ventilation–perfusion
15. carbonic anhydrase

B. Matching

1. E	3. B	5. A	7. T	9. M
2. G	4. U	6. J	8. X	10. K

C. True or False

The following items are false for the reasons stated. The other five are true.

1. Although O_2 and CO compete for the same binding site on hemoglobin, one hemoglobin molecule has four of these binding sites and could, therefore, carry O_2 on some and CO on others.
2. Atelectasis can be caused by other factors, such as blockage of the airway so air cannot get into a segment of the lung.
3. The body has very little means of monitoring and responding to the blood P_{O_2}; the stimulus to breathe comes mainly from CO_2.
4. Usually, hemoglobin gives up only about 22% of its oxygen on one pass through a systemic capillary (the utilization coefficient).
9. They are major COPDs but are caused primarily by cigarette smoking.

D. Multiple Choice

1. d	6. b	11. d	16. c	21. c
2. d	7. b	12. e	17. b	22. e
3. c	8. a	13. c	18. a	23. b
4. e	9. a	14. b	19. e	24. d
5. c	10. e	15. a	20. b	25. c

E. Word Origins and Meanings

1. No, seashell
2. No, ring
3. Yes
4. No, back of tongue
5. Yes
6. Yes
7. No, breath
8. Yes
9. No, air or breathing
10. Yes
11. painful, abnormal, difficult; dyspnea
12. inflamed; emphysema
13. tumor; adenocarcinoma
14. ladle; arytenoid
15. funnel; choanae

F. Which One Does Not Belong?

1. (c) is not capable of gas exchange, as the others are
2. (a) are membranes surrounding lungs; air passes through all the other structures
3. (c) is a muscle used in exhaling; others are muscles used during inspiration
4. (d) is part of the nose; others are parts of the larynx
5. (b) does not require maximum effort in order to measure; others do

G. Figure Exercise

1. A, B, C
2. K
3. C
4. Q
5. C
6. I
7. A
8. P
9. B, C, R
10. K, L, N

Chapter 23: The Urinary System

A. Short Answer

1. osmolarity
2. urea, uric acid
3. renal cortex
4. cortical radiate arteries, afferent arteriole
5. glomerulus
6. PTH/parathyroid hormone
7. proximal convoluted tubule
8. microvilli
9. transport maximum
10. nephroptosis
11. nephron loop
12. juxtaglomerular, glomerular filtration
13. renin, angiotensin II
14. sodium
15. internal urinary, external urinary

B. Matching

1. R	3. O	5. N	7. I	9. X
2. F	4. S	6. V	8. T	10. K

C. True or False

The following items are false for the reasons stated. The other five are true.

2. Aldosterone is secreted by the adrenal cortex.
3. The kidney clears only about 48% of the urea from the blood. Blood leaving the kidney has slightly more than half as much urea as the blood going into it.
6. No, 400 mL/day, called the obligatory water loss, is only the amount of urine that must be excreted to prevent azotemia. In oliguria and anuria, however, urine output may be much less than this.
9. This is wrong for two reasons—it overlooks the difference in molecular weights of the two chemicals (and therefore a great difference in the number of molecules per volume of a 2% solution of each), and it overlooks the fact that NaCl ionizes in water and sucrose does not.
10. Contraction of the detrusor is involuntary. Voluntary control of micturition is achieved by means of the external urinary sphincter.

D. Multiple Choice

1. b	6. c	11. c	16. b	21. a
2. a	7. c	12. b	17. e	22. c
3. a	8. d	13. e	18. c	23. d
4. b	9. a	14. e	19. b	24. e
5. e	10. a	15. a	20. a	25. c

E. Word Origins and Meanings

1. No, kidney
2. No, cup
3. Yes
4. Yes
5. No, foot
6. Yes
7. Yes
8. No, little or scanty
9. Yes
10. No, tasteless
11. crushing; lithotrypsy
12. vessel; mesangial
13. next to; juxtaglomerular
14. honey, sweet; diabetes mellitus
15. to urinate; micturition

F. Which One Does Not Belong?

1. (c) acts in direct contrast to the others, which all increase blood volume and blood pressure and decrease urine volume; however, (a) is the only steroid—the others are proteins
2. (d) are associated with the afferent and efferent arteriole and glomerular capillaries; the others are all part of the filtration barrier
3. (d) is made in the liver; the others are made in the kidney
4. (a) receives tubular fluid from several nephrons; the others are parts of a single nephron
5. (a) while it affects the other systems, it does not carry out excretion, as the others do

G. Figure Exercise

1. E
2. B, C
3. I
4. J, K, L
5. J, K, L
6. M
7. L, M, N
8. N
9. J, N
10. H
11. I
12. M
13. I
14. M
15. K, L

Chapter 24: Water, Electrolyte, and Acid–Base Balance

A. Short Answer

1. insensible
2. hyponatremia
3. phosphate
4. alkalosis
5. uncompensated
6. osmoreceptors
7. weak acid
8. fluid compartments
9. cutaneous transpiration
10. hydrogen ions (H^+)
11. hypotonic hydration
12. aldosterone
13. buffer system
14. ammonia
15. antidiuretic hormone

B. Matching

1. U	3. V	5. A	7. D	9. I
2. O	4. K	6. J	8. M	10. W

C. True or False

The following items are false for the reasons stated. The other five are true.

3. Although aldosterone does promote Na^+ reabsorption, a proportionate amount of water is reabsorbed along with it, and the concentration of the ECF does not change.
4. Chronic vomiting expels stomach acid from the body, leaving the body with excess base and causing alkalosis, not acidosis.
5. Acidosis and alkalosis are defined relative to the average pH of the ECF, 7.40. A pH of 7.2 is below (on the acidic side of) this, so it is considered to be acidosis.
7. Don't forget metabolic water.
10. There are no active transport pumps or other membrane carriers for water. Water is absorbed only by osmosis.

D. Multiple Choice

1. e	6. c	11. b	16. a	21. a
2. b	7. d	12. c	17. d	22. c
3. d	8. b	13. c	18. b	23. a
4. b	9. a	14. b	19. b	24. b
5. d	10. d	15. a	20. c	25. d

E. Word Origins and Meanings

1. No, across	6. No, potassium	11. volume; hypovolemia
2. Yes	7. Yes	12. to breathe; transpiration
3. Yes	8. No, outside or external to	13. calcium; hypercalcemia
4. No, to isolate	9. No, condition	14. intestine; parenteral
5. Yes	10. Yes	15. nourishment; hyperalimentation

F. Which One Does Not Belong?

1. (b) is fluid contained within cells; others are components of the ECF
2. (b) causes alkalosis; others cause acidosis
3. (d) while (b) looks like the obvious choice because it is the only anion, K^+ is the only ion found in the ICF; the rest are ECF ions
4. (d) is not critical in homeostasis, and we tolerate a wide range of phosphate concentrations; others all stimulate aldosterone release
5. (c) can cause hypercalcemia; others can cause hypocalcemia

Chapter 25: The Digestive System

A. Short Answer

1.   occlusal
2.   amylase, lipase
3.   cardiac, pyloric
4.   enterohepatic circulation
5. K^+, HCO_3^-
6.   cephalic
7.   acid-resistant bacteria (*Helicobacter*)
8.   hepatocytes
9.   trypsin
10. CCK
11. duodenocolic
12. pepsin
13. carboxypeptidase
14. chylomicrons
15. migrating motor complex

B. Matching

1. A
2. V
3. L
4. T
5. F
6. H
7. O
8. C
9. P
10. R

C. True or False

The following items are false for the reasons stated. The other five are true.

1.   Absorbed lipids travel in the lymphatic vessels to the left subclavian vein, thus getting into the bloodstream and circulating to all organs, not passing first through the liver.
5.   Intrinsic factor is used for absorption of vitamin B_{12}, not calcium.
6.   Some protein is absorbed by pinocytosis and may be responsible for some food allergies. The infant intestine, especially, absorbs IgA from the colostrum and milk.
7.   Chief cells are found in the gastric glands of the stomach, not in the intestinal crypts.
8.   HCl in the duodenum stimulates the release of secretin, which in turn stimulates the secretion of sodium bicarbonate by the pancreas. It's the sodium bicarbonate that neutralizes HCl.

D. Multiple Choice

1. d
2. c
3. a
4. e
5. b
6. d
7. a
8. d
9. a
10. c
11. e
12. c
13. d
14. a
15. d
16. c
17. b
18. e
19. a
20. d
21. b
22. e
23. e
24. b
25. e

E. Word Origins and Meanings

1. No, food
2. Yes
3. No, next to
4. No, intestine
5. Yes
6. No, sickle
7. Yes
8. Yes
9. Yes
10. No, juice
11. to draw; haustrum
12. empty, dry; jejunum
13. vomiting; emetic
14. gatekeeper; pylorus, pyloric
15. tail; caudate lobe

F. Which One Does Not Belong?

1. (b) is an accessory gland/organ; others are hollow regions of the alimentary canal through which chyme passes
2. (d) is a hormone from the duodenum; the others are involved with production of HCl, a process stimulated by gastrin
3. (a) it is the parasympathetic division of the ANS that makes up the enteric system of which (b) and (d) are part
4. (b) are pouches in the colon wall; others are associated with lipid digestion and absorption
5. (b) is an intrinsic salivary gland and produces lingual lipase and lysozyme; others are extrinsic salivary glands and have discrete ducts that empty mucus, amylase, and electrolytes into the oral cavity

G. Figure Exercise

1. G
2. Q
3. B, C, Q
4. P
5. D, E, N
6. K
7. Q
8. A, O, P
9. J, S, T
10. J, S, T
11. R
12. W
13. K
14. M
15. H

Chapter 26: Nutrition and Metabolism

A. Short Answer

1. anorexia, feeding center
2. hypoglycemia
3. insulin, epinephrine, glucagon
4. arcuate nucleus (hypothalamus)
5. high-density lipoproteins
6. complete, essential
7. negative nitrogen balance
8. absorptive
9. metabolic rate (TMR)
10. pyruvic acid
11. oxaloacetic acid
12. sympathetic, glucagon
13. chemiosmotic
14. glycemic index
15. heat exhaustion

B. Matching

1. R
2. D
3. H
4. E
5. V
6. K
7. Q
8. B
9. P
10. G

C. True or False

The following items are false for the reasons stated. The other five are true.
1. Water is a nutrient because it is absorbed and made a part of the body's tissues.
2. Cellulose is not a nutrient because it is never absorbed or made a part of the body.
3. Anaerobic fermentation is a way of keeping glycolysis going. Glycolysis produces a small amount of ATP, but it does so in the cytosol, not in the mitochondria.
6. Glycogenesis removes glucose from circulation to synthesize glycogen; glycogenolysis is the breakdown of glycogen
8. Most ATP is produced by the mitochondrial electron transport chain.

D. Multiple Choice

1. b
2. d
3. e
4. c
5. e
6. c
7. e
8. a
9. e
10. c
11. a
12. d
13. e
14. e
15. c
16. b
17. d
18. c
19. e
20. a
21. a
22. e
23. b
24. a
25. c

E. Word Origins and Meanings

1. No, appetite
2. Yes
3. No, sugar or glucose
4. No, splitting
5. Yes
6. No, push or thrust
7. Yes
8. Yes
9. Yes
10. Yes
11. bag; ascites
12. thin; leptin
13. like, resembling; ascites
14. without; anorexia
15. vomiting; hematemesis

F. Which One Does Not Belong?

1. (b) is a glycolysis intermediate; others are involved in the Krebs/citric acid cycle
2. (b) is a water-soluble B-vitamin; others are fat-soluble vitamins
3. (c) produces no ATP; the other processes produce varying amounts of ATP
4. (a) is a reduced coenzyme; others are components of the electron transport chain
5. (a) functions as a cofactor in metabolic reactions; others are antioxidants

Chapter 27: The Male Reproductive System

A. Short Answer

1. testis-determining factor
2. homologous
3. scrotum, gubernaculum
4. perineum
5. seminiferous tubules
6. male parent (Y chromosome)
7. pampiniform, countercurrent heat
8. ejaculatory duct
9. testosterone
10. lacunae
11. androgens (primarily testosterone)
12. testosterone, libido
13. crossing over
14. chryptorchidism
15. bulbospongiosus

B. Matching

1. V	3. G	5. L	7. C	9. H
2. F	4. S	6. O	8. N	10. A

C. True or False

The following items are false for the reasons stated. The other five are true.

1. A haploid cell has 23 *unpaired* chromosomes and a diploid cell has 23 *pairs* (46 chromosomes total).
2. Meiosis I, not II, converts a germ cell from diploid to haploid.
5. The conversion of spermatids to spermatozoa (spermiogenesis) involves no cell divisions at all.
6. Drugs such as tadalafil and sildenafil act by inhibiting phosphodiesterage, which breaks down cGMP, and, therefore, prolong erection.
9. Semen consists mostly of secretions of the prostate and seminal vesicles. If sperm are absent (as in vasectomized males), there is very little difference in the volume of the semen.

D. Multiple Choice

1. c	6. d	11. c	16. b	21. c
2. e	7. a	12. a	17. a	22. d
3. d	8. c	13. e	18. a	23. e
4. b	9. b	14. d	19. c	24. e
5. a	10. b	15. b	20. a	25. b

E. Word Origins and Meanings

1. Yes
2. No, hidden
3. No, testis
4. No, sheath
5. Yes
6. No, seam
7. Yes
8. Yes
9. No, tip
10. Yes
11. coil, helix; helicine arteries
12. steer, guide; gubernaculum
13. to creep; herpes
14. network; rete testis
15. grow up; puberty

F. Which One Does Not Belong?

1. (b) are diploid cells that undergo mitosis; others are haploid cells produced during meiosis
2. (d) are primary sex organs; others are secondary sex organs
3. (b) is a viral STD; others are caused by bacteria
4. (d) shrinks in order for the testes to descend into the scrotum; others are structures within the testes
5. (c) is not a function of testosterone, as the others are

G. Figure Exercise

1. Q
2. C, D, E, G, H, K, M
3. L
4. E
5. Q, R
6. G
7. S
8. C, D, H
9. B, L
10. S

Chapter 28: The Female Reproductive System

A. Short Answer

1. follicle, ovulation
2. uterine (fallopian) tube
3. myometrium
4. endometrium
5. vulva (pudendum)
6. glans clitoris
7. lactiferous sinus
8. myoepithelial, oxytocin
9. menopause, climacteric
10. pubarche
11. primary oocyte, secondary oocyte
12. decreases, atresia
13. zona pellucida, corona radiata
14. corpus luteum
15. prolactin

B. Matching

1. R
2. F
3. C
4. M
5. L
6. H
7. U
8. K
9. O
10. I

C. True or False

The following items are false for the reasons stated. The other five are true.
1. Only the oocyte and some of the cumulus oophorus cells, not the follicle, are released by ovulation.
4. HCG is secreted by the blastocyst and later by the chorion, not by the ovary.
5. The vestibular bulbs are subcutaneous erectile tissues on each side of the vagina; the erectile tissues of the clitoris are corpora cavernosa.
6. Cow's milk contains *too much* protein—about three times as much as human milk.
9. The follicles for a new cycle begin to develop about two months before menstruation begins.

D. Multiple Choice

1. b
2. d
3. c
4. a
5. e
6. a
7. e
8. c
9. a
10. e
11. b
12. d
13. d
14. b
15. e
16. c
17. d
18. d
19. b
20. a
21. b
22. e
23. a
24. e
25. a

E. Word Origins and Meanings

1. Yes
2. No, uterus
3. Yes
4. No, uterus
5. Yes
6. No, desire
7. No, month
8. Yes
9. No, egg
10. Yes
11. new; neonatal
12. pregnancy; gestation
13. child; puerperium
14. pain; mittelschmerz
15. yellow; corpus luteum

F. Which One Does Not Belong?

1. (d) is not a normal occurrence during pregnancy; the others do occur
2. (b) is made in the hypothalamus; the others are secreted by the ovary
3. (a) does not occur during ovulation, as the others do
4. (c) is a barrier method of contraception; the others prevent implantation or ovulation
5. (b) increases in pregnancy to increase maternal blood volume; the others are correct hormone responses during pregnancy

G. Figure Exercise

1. D
2. T (O, P, Q)
3. P
4. V
5. T (O, P, Q)
6. A
7. K
8. F
9. N, T (O, P, Q)
10. I
11. M
12. B
13. V
14. N
15. B

Chapter 29: Human Development and Aging

A. Short Answer

1. capacitation
2. cortical granules
3. adolescence
4. monozygotic
5. placenta
6. nondisjunction
7. life span
8. photoaging
9. syncytiotrophoblast
10. Down syndrome
11. amniotic fluid
12. vertex
13. ductus arteriosus
14. zygote
15. hypothalamus

B. Matching

1. B	3. L	5. S	7. E	9. R
2. M	4. V	6. H	8. A	10. N

C. True or False

The following items are false for the reasons stated. The other five are true.

1. Eggs are viable for only about 24 hours; it takes 72 hours for an egg to travel the uterine tube.
3. The yolk sac is the source of the embryo's first blood and germ cells.
8. Elderly people have less subcutaneous fat than younger people have.
9. Fertilization occurs in the distal end of the uterine tube, not in the uterus.
10. There seems to be little prospect for increasing the life span much beyond its present limit.

D. Multiple Choice

1. b	6. d	11. d	16. a	21. e
2. b	7. e	12. e	17. c	22. b
3. a	8. a	13. a	18. b	23. e
4. c	9. b	14. b	19. a	24. c
5. d	10. c	15. b	20. d	25. c

E. Word Origins and Meanings

1. No, bud or precursor
2. Yes
3. Yes
4. No, yellow
5. Yes
6. No, birth or born
7. No, monster
8. Yes
9. Yes
10. No, end
11. old age; progeria
12. with; congenital
13. mulberry; morula
14. rule, regularity; anomaly
15. together; syncytiotrophoblast

F. Which One Does Not Belong?

1. (a) does not appear to decrease with age, as the others do
2. (a) is not an embryonic membrane, as the others are, but contains umbilical arteries and umbilical vein
3. (b) is an infectious disease that can cause congenital anomalies; the others are chemical teratogens
4. (d) is found in a neonate; others are shunts found in fetal circulation
5. (d) is formed from embryonic ectoderm; others arise from mesoderm

G. Identification Exercise

1. decrease
2. increase
3. decrease
4. no change
5. increase
6. increase
7. decrease
8. decrease
9. decrease
10. increase
11. increase
12. decrease
13. increase
14. decrease
15. increase
16. decrease
17. decrease
18. increase
19. increase
20. little change

Notes

Notes

Notes

Notes